Springer Atmospheric Sciences

More information about this series at http://www.springer.com/series/10176

Indrani Roy

Climate Variability and Sunspot Activity

Analysis of the Solar Influence on Climate

 Springer

Indrani Roy
Mathematics and Physical Sciences
University of Exeter, College of Engineering
Exeter, UK

ISSN 2194-5217   ISSN 2194-5225   (electronic)
Springer Atmospheric Sciences
ISBN 978-3-030-08372-4   ISBN 978-3-319-77107-6   (eBook)
https://doi.org/10.1007/978-3-319-77107-6

This Springer imprint is published by the registered company Springer International Publishing AG part
of Springer Nature.
The registered company address is: Gewerbestrasse 11, 6330 Cham, Switzerland

Dedicated to my late parents
Sri Barin Kar
&
Smt Jyotsna Kar

Contents

Part II Atmosphere-Ocean Coupling and Solar Variability

Sources to Figures

Fig. 7.7   http://data.giss.nasa.gov/gistemp/

Fig. 13.1   http://www.climate4you.com/Polartemperatures.htm#Arctic   monthly
             surface air temperatures north of 70°N

Fig. 13.2   http://psc.apl.washington.edu/wordpress/wp-content/uploads/schwei-
             ger/ice_volume/BPIOMASIceVolumeAnomalyCurrentV2.1.png

Fig. 13.3   http://scitechdaily.com/images/Study-Shows-Volume-of-Arctic-Sea-
             Ice-Has-Increased.jpg

Fig. 13.4   EUMETSAT, OSI SAF (http://osisaf.met.no, graph plotted on 16/12/14.),
             also in http://kaltesonne.de/wp-content/uploads/2014/12/arktis2.gif

Fig. 13.5   http://www7320.nrlssc.navy.mil/hycomARC/navo/arcticicen_nowcast_
             anim365d.gif

Fig. 13.8   http://www.vencoreweather.com/blog/2016/4/11/215-pm-global-sea-
             ice-makes-a-strong-comeback, link on 11/4/16

Fig. 13.9   http://eoimages.gsfc.nasa.gov/images/imagerecords/8000/8239/antarc-
             tica_avhrr_81-07_lrg.pdf, link accessed on 2/4/18

Fig. 14.2   https://ktwop.files.wordpress.com/2013/10/73-climate-models_reality.
             gif, uploaded on 30/12/2017. Credit: JR Christy, University of Alabama,
             Huntsville-model output from KNMI

Fig. 14.3   http://www.cato.org/blog/current-wisdom-record-global-temperature-
             conflicting-reports-contrasting-implications.   uploaded   on   30/12/17,
             Cedit: © The Cato Institute 2014. Used by permission

Fig. 15.1   http://en.wikipedia.org/wiki/File:EM_spectrum.svg,   uploaded   on
             30/12/17

Fig. 15.2   https://commons.wikimedia.org/wiki/File:Atmospheric_Transmission.
             png. CC BY-SA 3.0, image loaded on 30/12/2017

Fig. 15.3   earthobservatory.nasa.gov/Features/EnergyBalance/page7.php.
             uploaded on 30/12/2017, credit NASA's Earth Observatory

Fig. 15.4   Carbon Connections. http://carbonconnections.bscs.org. Copyright ©
             BSCS. All rights reserved. Used with permission

Fig. 15.5   https://commons.wikimedia.org/w/index.php?curid=10684392,   loaded
             on 19/12/2017. (By Vostok-ice-core-petit.png: NOAA derivative work:
             Autopilot (Vostok-ice-core-petit.png) [CC-BY-SA-3.0 (http://creative-
             commons.org/licenses/by-sa/3.0/)   or   GFDL   (http://www.gnu.org/
             copyleft/fdl.html)], via Wikimedia Commons.)

Fig. 15.6   https://www.frontiersin.org/files/Articles/317793/feart-05-00104-
             HTML/image_m/feart-05-00104-g002.jpg,   uploaded   on   19/12/2017,
             Copyright © 2017 Lüning and Vahrenholt

Fig. 16.1   http://www.carbonbrief.org/in-brief-how-much-do-volcanoes-influ-
             ence-the-climate uploaded on 30/12/17, used with permission

Fig. 17.1   http://www.environment.gov.au/protection/ozone/ozone-science/ozone-
             layer/antarctic-ozone-hole, Animation link (dt 30/12/17)

Fig. 18.2   https://commons.wikimedia.org/wiki/File%3APIA16938-RadiationSources-InterplanetarySpace.jpg, uploaded on 30/12/17, picture by NASA/JPL-Caltech/SwRI (http://photojournal.jpl.nasa.gov/jpeg/PIA16938.jpg) [Public domain], via Wikimedia Commons

Fig. 18.3   http://www.climate4you.com/Sun.htm# Cosmic ray intensity and sunspot activity, uploaded on 30/12/2017, credit Germany Cosmic Ray Monitor in Kiel (GCRM) and NOAA's National Geophysical Data Center (NGDC)

Introduction

The energy to drive the global atmospheric circulation originates from the Sun. During higher solar activity, the Earth is subjected to enhanced solar irradiance. However, the connection between the Sun and climate has been a puzzle. This is because during a typical 11-year cycle the solar energy output varies by only about 0.1% (Lean and Rind 2001). Based on energy considerations, such a change is too small to produce significant changes in surface conditions. The degree to which variation in solar output affects climate has been the topic of extensive research for a long time.

But nowadays, there is a common consensus that the variations in the UV part of the spectrum between solar maxima and minima (6–8%) leads to more ozone and warming during solar maxima in the upper stratosphere (Crooks and Gray 2005; Hood 2004 and Haigh 1994). Such modulations can impact tropospheric climate. The primary motivations for exploring the areas of solar 11-year cycle variability on climate are the following:

- The Sun is the principal source of energy in the earth; it causes day/night and seasons. There are reasons to believe that solar variability can influence the climate.
- As the Sun follows the 11-year cyclic variability, the knowledge about the Sun–climate relationship on that scale can be used for future climate prediction purpose.
- However, regarding energy output, there is only 0.1% change from solar minimum to maximum years of the 11-year variability, which is too negligible to influence the climate of the earth.
- Moreover, a significant signal is detected on some meteorological parameters, but most of the time that is regionally different. It can also vary with time periods. Unless there are mechanisms to support such behaviour, it may be a mere coincidence.

Other solar-related drivers are also found to influence the climate of earth, for example, geomagnetic activity, total solar irradiance (TSI) and solar cosmic or magnetospheric energetic particle precipitation, and are discussed briefly in the final chapter of the book. However, the current analysis mainly focuses on solar 11-year cyclic variability (as detected for Sunspot number (SSN)). It is because the knowledge about the Sun–climate relationship on that scale can be used for future projection purposes and hence has implications for improved climate prediction. Moreover, unlike other solar-related drivers, there is a very well-accepted mechanism for the SSN, based on solar UV-related variability.

This work is the collection of lecture notes as well as synthesised analyses of published papers on the described subjects. It is divided into three parts: Part I discusses general circulation, climate variability, stratosphere–troposphere coupling and various teleconnections. Part II mainly explores the area of different solar influences on climate and also discusses about ocean –atmosphere coupling. But without a prior knowledge of other important influences on the earth's climate, the understanding of the actual role of the Sun remains incomplete. Hence, Part III covers burning issues such as greenhouse gas warming, volcanic influences, ozone depletion in the stratosphere and Arctic and Antarctic sea ice. At the end of the book, there are few questions and exercises for students.

References

Lean J, Rind D (2001) Earth's response to a variable Sun. Science 292(5515):234–236

Hood LL (2004) Effects of solar UV variability on the stratosphere, solar variability and its effects on climate. Geophys Monogr Amer Geophys Union 141:283–303

Crooks SA, Gray LJ (2005) Characterisation of the 11-year solar signal using a multiple regression analysis of the ERA-40 dataset. J Clim 18(7):996–1015. https://doi.org/10.1175/JCLI-3308.1

Haigh JD (1994) The role of stratospheric ozone in modulating the solar radiative forcing of climate. Nature 370:544–546

Part I
Climatology, General Circulation, Climate Variability and Stratosphere-Troposphere Coupling

Abstract Part I initially focused on basic definitions of Climatology, General Circulation, Climate Variability and Stratosphere Troposphere Coupling. There was a discussion on Climatology of Sea Level Pressure (SLP) and Sea Surface Temperature (SST) which was followed by defining Hadley and Walker circulation. In the subsequent chapter, major modes of climate variability are included with their spatial characteristic and temporal behavior. It is then followed by an overview of Stratosphere-Troposphere coupling. Later on, there is a discussion based on teleconnection among various modes. Discussion on how robust solar influences around different places are detected is included afterwards. Finally, there is a discussion on Total Solar Irradiance (TSI) and how it is reconstructed.

Keywords Climatology · Modes of variability · Stratosphere-Troposphere Coupling · Total Solar Irradiance (TSI) · General circulation

Chapter 1
Climatology and General Circulation

Abstract This chapter focused on basic definitions of climatology and general circulation. There was a discussion on climatology of sea level pressure (SLP) and sea surface temperature (SST) which was followed by defining north-south Hadley and east-west Walker circulation. It also defined and described Ferrel cell, Polar cell and various jets. It explained thermal-wind balance relationship and its relevance to jet formation.

Keywords Climatology · Sea Level Pressure (SLP) · Sea Surface Temperature (SST) · Hadley Cell · Ferrel Cell · Polar Cell · Walker Circulation · Subtropical Jet · Thermal Wind Balance · Aleutian Low · Icelandic Low · Azore High · Intertropical Convergence Zone (ITCZ)

To begin with, I start with the definition of weather and climate. Weather is the changing atmospheric conditions (as they affect people), such as precipitation, mist, fog, etc. or meteorological elements like temperature, humidity, winds, etc. On the other hand, the climate is weather averaged over an extended period (say, 30 years).

Below is a brief description about climatology (30 years average) of sea surface temperature (SST) and sea level pressure (SLP) followed by a description of the general circulation.

1.1 Climatology: SLP and SST

The climatology of SLP (Fig. 1.1) and SST (Fig. 1.2) is discussed during two different seasons, the boreal winter and summer. Boreal winter or summer means Northern Hemispheric (NH) winter or summer season. In contrast, austral winter or summer means Southern Hemispheric (SH) winter or summer.

The mean SLP during boreal winter (December-January-February, DJF) differed to that from boreal summer (June-July-August, JJA) and illustrated in Fig. 1.1a, b, respectively. The seasonal variations in SLP are most apparent in the NH. During winter the high-latitude oceans are characterised by low-pressure centres with the Aleutian Low (AL) and Icelandic Low centre in the northern margins of the Pacific

I. Roy, *Climate Variability and Sunspot Activity*, Springer Atmospheric Sciences, https://doi.org/10.1007/978-3-319-77107-6_1

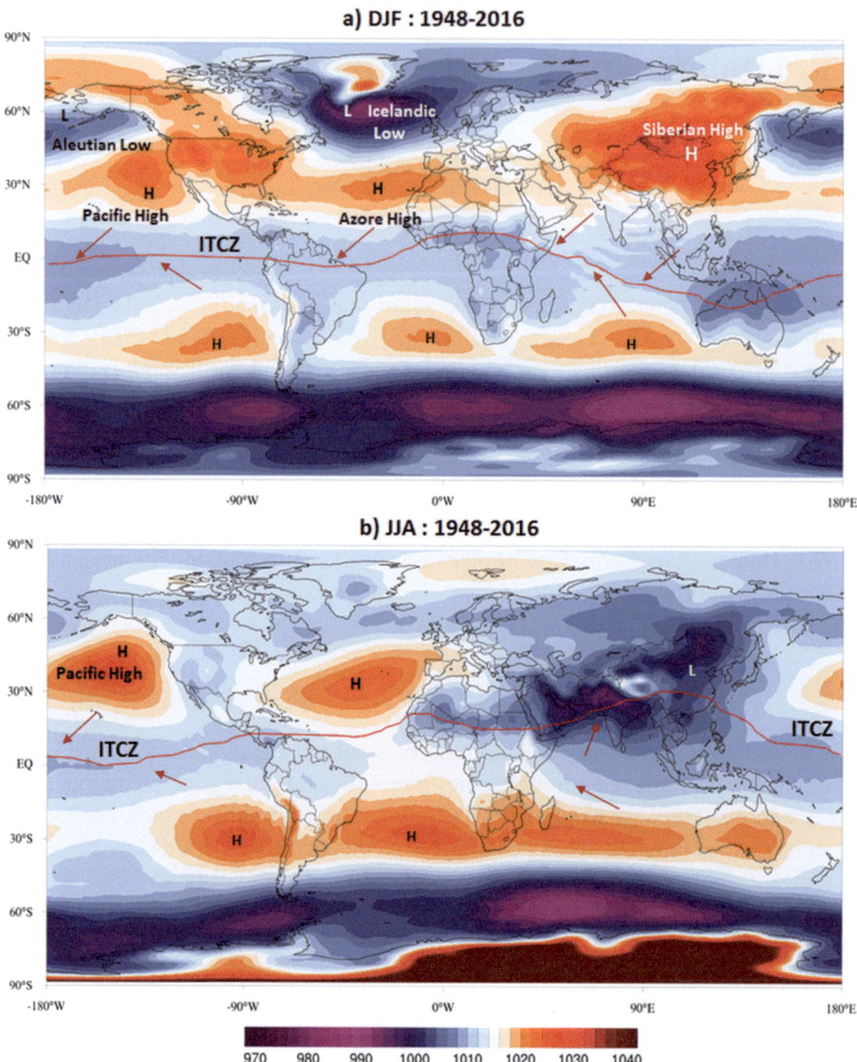

Fig. 1.1 Mean SLP(mb) during (**a**) northern winter (DJF) and (**b**) northern summer (JJA). (Source: Background maps generated via http://cci-reanalyzer.org/, using NCEP/NCAR Reanalyses v1 data, ClimateReanalyzer.org, University of Maine, Climate Change Institute)

and Atlantic Oceans, respectively, whereas a high-pressure centre lies over Asia. During summer, the land-sea pressure contrast is reversed in midlatitudes, with the highest pressures over the oceans and the lowest pressures over the land areas. It is seen from that figure that most high-pressure regions persist throughout the year. Although, the Pacific High (PH) and Azore High are weaker in the winter than summer. The dominant low-pressure feature during NH summer is centred over Asia at about 30°N and associated with the Asian summer monsoon. Movement of the intertropical convergence zone (ITCZ), further north around the Indian Ocean

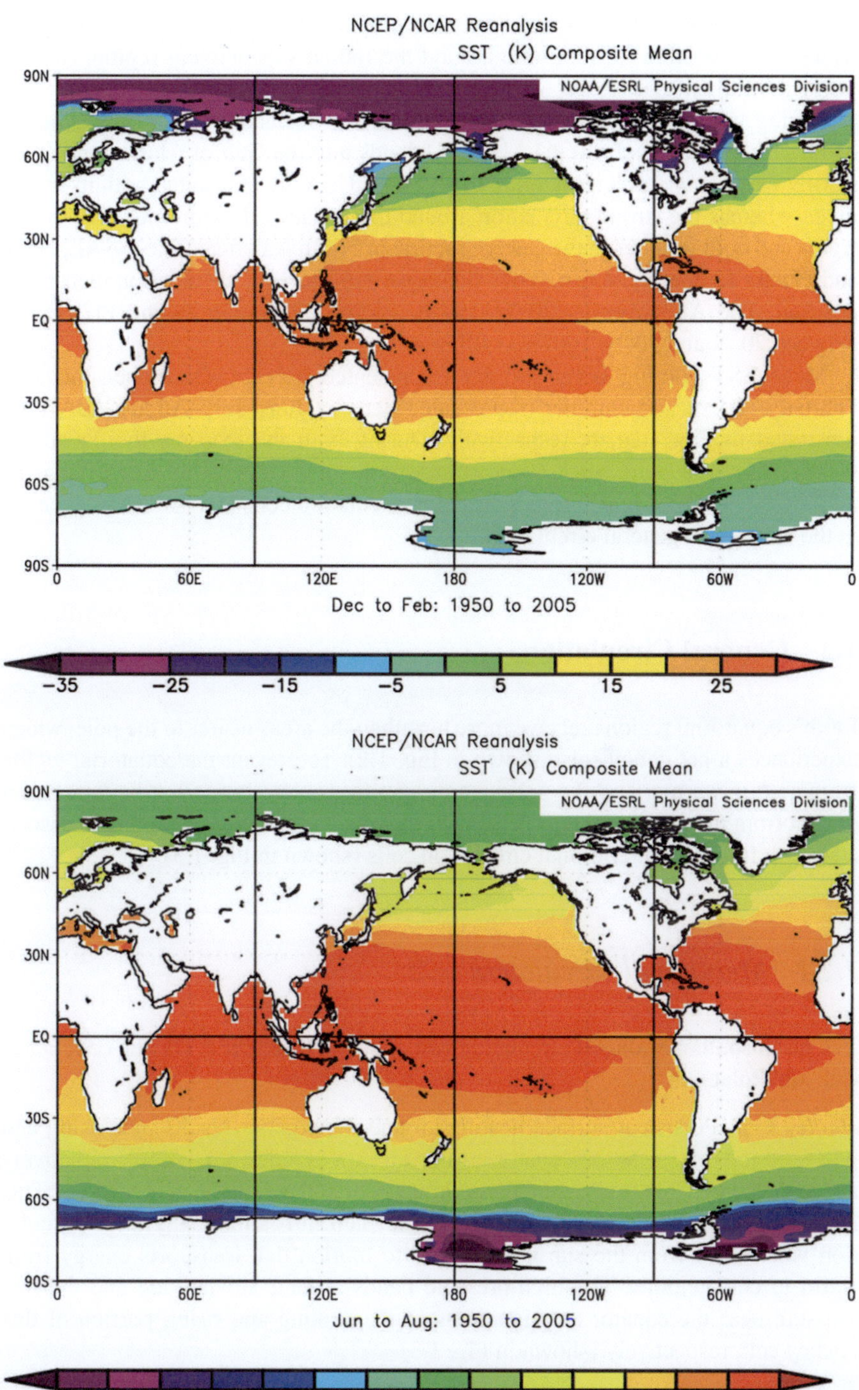

Fig. 1.2 NOAA extended SST (°C) composite mean. (Source: Plots generated using the data from NOAA/OAR/ESRL PSD, Boulder, Colorado, USA, from their website at (http://www.esrl.noaa.gov/psd/))

region during summer, (shown in Fig. 1.1b compared to 1.1a) is clearly noticeable – which is responsible for monsoon around the Indian subcontinent region, causing heavy rainfall.

During winter (shown in Fig 1.1a), the AL and Icelandic Low are well developed. The AL extends from the Aleutian Islands into the Gulf of Alaska, and much stormy weather and precipitation in the Western USA are associated with its movement, whereas the strong circulation around the Icelandic Low produces northerly winds and cold weather in the eastern section of North America. Like the AL, ITCZ and PH are frequently in use in our subsequent discussion; we mention their mean position. The AL covers ~120°W–130°E, 35°N–70°N, whereas the PH centres between 20°N and 50°N, 100°W–140°E.

Mean SSTs during boreal winter (represented here by December–January–February (DJF)) and summer (represented here by June–July–August (JJA)) are illustrated in Fig. 1.2a, b, respectively. During both the seasons, the equator is warmer than the pole. Such temperature gradient is responsible for driving the meridional heat transport through different circulation cells and is described below in the chapter on general circulation.

1.2 General Circulation

Earth's equatorial regions receive more heat than the areas nearer to the pole, which experiences a net deficit (also shown in Fig. 1.2). To prevent the equatorial region from getting warmer and the polar region getting cooler, there must be a transport of heat from the equatorial region towards the pole. Such transport of heat is associated with following meridional circulation cells (shown in Fig. 1.3).

1.2.1 Meridional Circulation

The north south meridional circulation comprises of three cells; Hadley cell, Ferrel Cell and Polar cell.

Hadley Cell Persistent surface heating around the equator results in a rising and poleward moving air at the equator; as this air moves poleward, it cools radiatively and sinks near 30° latitude; it finally returns towards the equator, at low levels. This meridional circulation cell is called the Hadley cell. It is a thermally direct circulation where heat from the sun is converted to motion that transports energy from warm to cold regions. The high-pressure bands around 30° latitude and the low pressure near the equator are signatures of the sinking and rising portion of this Hadley cell, respectively (shown in Fig. 1.3).

Ferrel Cell In midlatitudes the cell that circulates in the opposite direction to the Hadley cell is known as the Ferrel cell, and the overall movement of surface air in this cell is from 30° latitude to 60° latitude (shown in Fig. 1.3). This cell sinks over warm temperature zone and rises over cold temperature zone. This cell is not driven by thermal forcing but generated by eddy forcing (and associated weather systems). They are the deviations from the time or zonal average and are a key component of the general circulation of the atmosphere. Around midlatitudes, the biggest contribution to meridional energy transport comes from eddies in the air, and they transport heat poleward. Midlatitude cyclones and anticyclones are the major transient eddies that serve a significant part in meridional transports of moisture, momentum and heat. Those weather systems are generated from the baroclinic instability which is associated with large south-north temperature gradients. Cyclones around midlatitude are marked by well-defined fronts that segregate the cold air mass from the north to that from warm air mass from the south. Those are different from tropical hurricanes, which do not have frontal features. The characteristic time scale of midlatitude cyclones is 7–10 days and has typical spatial scales of wavenumbers 5–6.

Polar Cell Air circulates within the troposphere, limited vertically by the tropopause at about 8 km in higher latitudes. In both hemispheres, warm air rises at around 60° latitude and moves poleward through the upper troposphere. When air reaches the polar area, being cooled considerably, it descends as a cold, dry high-pressure area, which moves away from the pole along the surface and generates the polar cell (shown in Fig. 1.3).

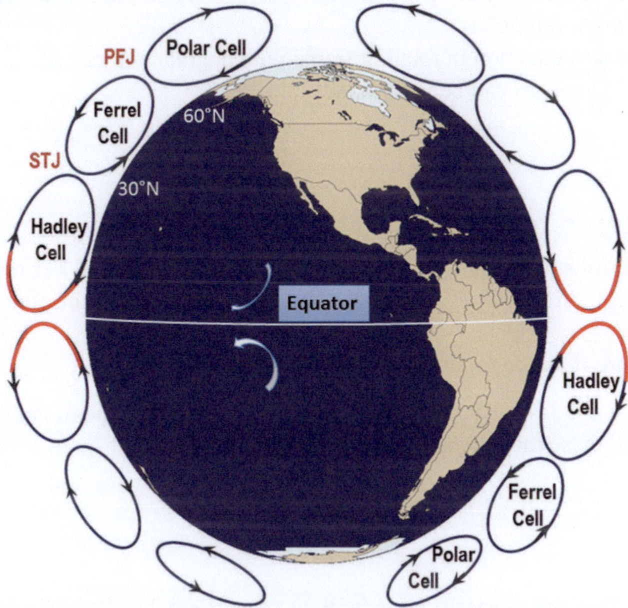

Fig. 1.3 Schematic representation of the general circulation showing positions of Hadley Cell, Ferrel Cell, Polar Cell, subtropical jets (STJ) and polar jets (PFJ). (Source: Background map generated via http://cci-reanalyzer.org/, ClimateReanalyzer.org, University of Maine, Climate Change Institute)

Jet streams are seen at approximate boundaries between these cells (shown in Fig. 1.3) and are formed around the upper tropospheric region (mainly around the tropopause region – the border between the stratosphere and troposphere) due to the **thermal wind balance relationship**. The subtropical jet (STJ) lies at the interface between the Hadley and the Ferrel cell, whereas the polar front jet (PFJ) also known as the midlatitude jet or the polar jet are observed between the Ferrel cell and the Polar cell.

Around 30°N there is sinking air which is typically free of substantial precipitation. Major deserts of the northern hemisphere are found here, e.g. Middle East, Sahara, etc.

1.2.2   Jet Formation: Thermal Wind Balance Relationship

It is derived from the equation of state and Navier-Stokes equation as shown below:

- *Equation of State*

$$PM = \rho RT$$

where T and P are the atmospheric temperature and pressure, respectively, ρ is the density, R is the universal gas constant, and M is the molecular weight.
- *Navier-Stokes Equation*
  Navier-Stokes equation in rotating frame with angular velocity Ω.
  After scale analysis can be written as:

$$f k \times U_h = -\frac{1}{\rho} \nabla_h P$$

where Coriolis parameter is f, $f = 2\Omega \sin \Phi$, Φ is latitude, and k is unit vector in the vertical direction. U is vector velocity, and h represents the horizontal direction.
- *Hydrostatic Balance Equation*

  $\partial P/\partial z = -g\rho$, z is vertical direction and g is acceleration due to gravity.
  Combined these three equations can be written as:

$$\frac{\partial U_h}{\partial z} \approx \frac{g}{fT} k \times \nabla_h T$$

It indicates that the westerly wind will increase with height when temperature decreases towards the pole. This is known as thermal wind balance relationship and responsible for jet formation.

1.2.3   Walker Circulation

Apart from north-south (meridional) circulations, as discussed, there is an west-east (zonal) circulation over the tropical Pacific Ocean, known as the Walker circulation, as described below.

The Walker circulation is a large-scale zonal flow, with air rising over the western Pacific and descending around the eastern Pacific as shown in Fig. 1.4. It is generated by the pressure gradient force resulting from a low-pressure system over Indonesia with a high-pressure system over the eastern Pacific Ocean (the mean position of this high- and low-pressure system is also shown in Fig. 1.1a). The Walker cell varies interannually, which is associated with a reversal in direction of winds (from Fig. 1.4, shown with arrow).

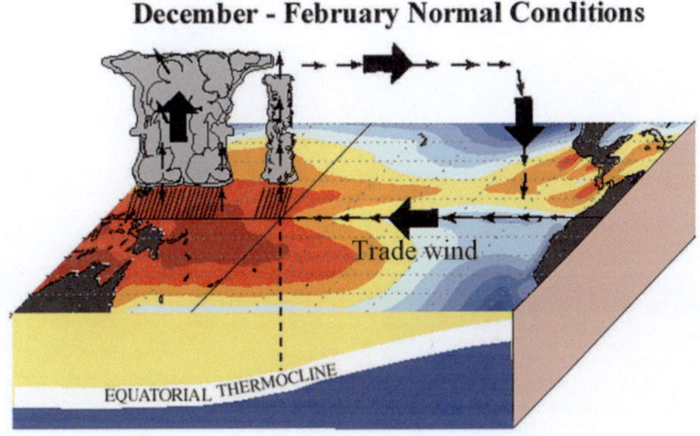

Fig. 1.4 Walker circulation shown with arrow. (Source: http://www.cpc.noaa.gov/products/analysis_monitoring/ensocycle/meanrain.shtml, link accessed on 10/12/2017, credit: National Oceanic and Atmospheric Administration (NOAA) Climate Prediction Center)

Chapter 2
Major Modes of Variability

Abstract This chapter focused on major modes of variability which serve the key role in controlling the regional climate. In terms of tropospheric variability, it defined and discussed ENSO (El Niño Southern Oscillation), NAO (North Atlantic Oscillation), AO and AAO (Arctic Oscillation, Antarctic Oscillation), Indian Monsoon, Indian Ocean Dipole (IOD), PDO (Pacific Decadal Oscillation) and AMO (Atlantic Multidecadal Oscillation). Later it attended stratosphere variability; this constitutes QBO (quasi-biennial oscillation) and SSW (stratospheric sudden warming). Main characteristic features of each of these modes were elaborately discussed.

Keywords Modes of variability · ENSO · NAO · AO · AAO · PDO · AMO · Indian Summer Monsoon (ISM) · Indian Ocean Dipole (IOD) · QBO · SSW

There are various modes of climate variability, which play important roles in determining the characteristic of different regions of the climate of the Earth as shown in Fig. 2.1

These are classified below as whether they are features of the troposphere or the stratosphere.

- Troposphere: ENSO (El Niño Southern Oscillation)
  NAO (North Atlantic Oscillation)
  AO and AAO (Arctic Oscillation, Antarctic Oscillation)
  PDO (Pacific Decadal Oscillation)
  AMO (Atlantic Multidecadal Oscillation)

- Stratosphere: QBO (Quasi-Biennial Oscillation)
  SSW (Stratospheric Sudden Warming)

© Springer International Publishing AG, part of Springer Nature 2018

I. Roy, *Climate Variability and Sunspot Activity*, Springer Atmospheric Sciences, https://doi.org/10.1007/978-3-319-77107-6_2

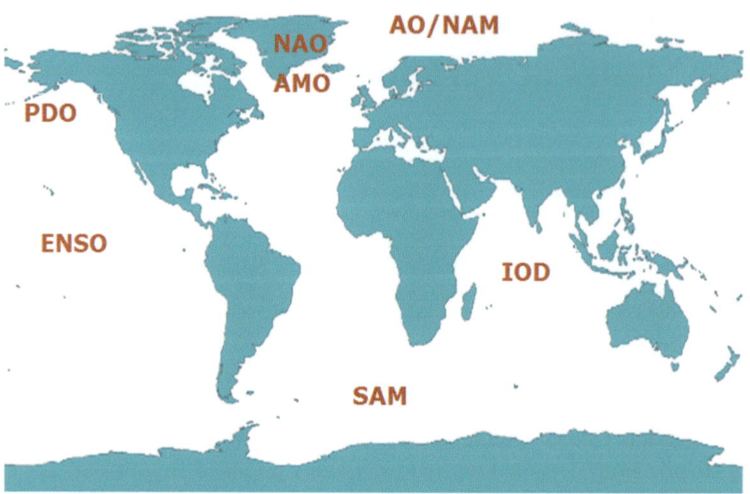

Fig. 2.1 Various modes ofclimate variabilityand the region of their influence

2.1   Variability in the Troposphere

2.1.1   *El Niño and Southern Oscillation (ENSO)*

ENSO is the leading mode of variability in the tropics, although its influence is felt globally. The Southern Oscillation (SO) is a large-scale fluctuation in air pressure between the western and the eastern tropical Pacific about the international dateline. Tahiti (17.6°S, 149.4°W) and Darwin in Australia (12.3°S, 130.5°E) are two opposite ends of this SO's seesaw, and the SO index is calculated based on the difference in air pressure between these two places (shown in Fig. 2.2a).

The El Niño is named for the periodic warming which occurs around the eastern tropical Pacific and which adversely affects the fishing industry around the coast of Peru and Ecuador. In general, a smoothed time series of the SO index corresponds very well with changes in ocean temperatures across the eastern tropical Pacific (Fig. 2.2b). These two interrelated phenomena, the El Niño and SO, abbreviated as the ENSO, control a large proportion of climate variability in the tropics.

The ENSO index time series is based on oceanic temperatures of the eastern tropical Pacific. Different formulations are in use, based on slight different geographical considerations. The most commonly known indices are Niño 1 + 2, Niño 3,Niño 3.4and Niño 4 as shown in Fig. 2.2c depicting their geographic coverage. Geographic coverage of various commonly used Niño regions are as follows:

- Niño 2 and 1: (0–10 °S, 90−80 °W)
- Niño 3: (5 N–5 °S, 150−90 °W)
- Niño 3.4: (5 N–5 °S, 170−120 °W)
- Niño 4: (5 N–5 °S, 160 °E −150 °W)

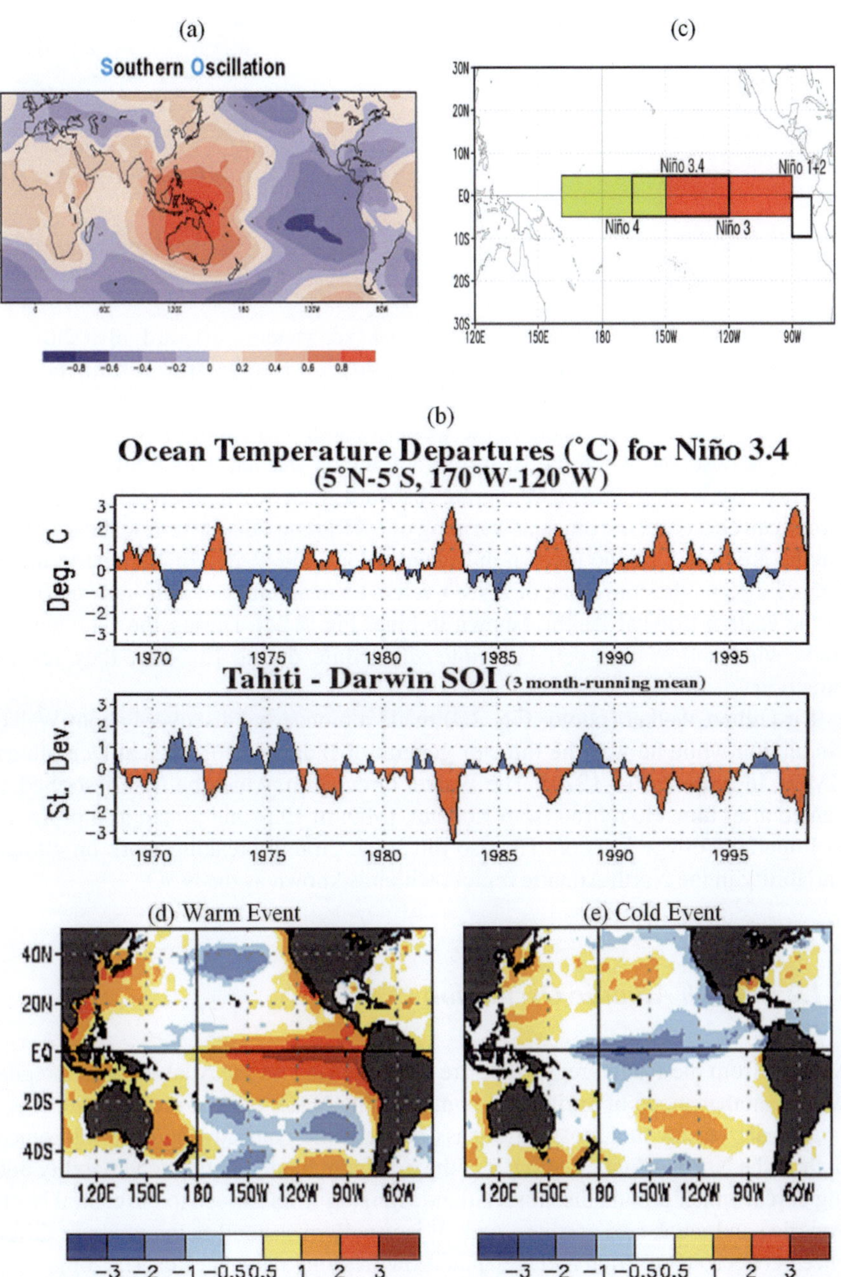

Fig. 2.2 Representation of ENSO: (**a**) SLP variation (hPa) showingSouthern Oscillation; (**b**) temperature of Niño 3.4 and Southern Oscillation anti-correlation; (**c**) different Niño regions; (**d**) departure of SST (°C) during DJF for warm events of ENSO; and (**e**) cold events of ENSO. (Source: http://www.cpc.noaa.gov/products/analysis_monitoring/ensocycle/ensocycle.shtml, Credit: National Oceanic and Atmospheric Administration (NOAA) Climate Prediction Center)

Fig. 2.3 Sir Gilbert Walker (1868–1958). (Source: https://pt. wikipedia.org/wiki/ Gilbert_Walker, link accessed on 10/12/2017)

In the tropical Pacific, trade winds drive surface waters westward (also shown in Fig. 1.4 to describe the Walker circulation). Surface water that travels from the eastern Pacific all the way extended to western Pacific (thus absorbing more solar radiation) become warmer reaching in the western Pacific. El Niño, the warm events of ENSO, is observed when the easterly trade winds weaken, which allows warmer waters of the western Pacific to migrate eastward. That eventually reaches the South American coast causing unusual warming of SST around the eastern seaboard of the Pacific and thus El Niño (shown in orange colour, Fig. 2.2d). In contrast to the El Niño, La Niña, the cold event of ENSO, refers to an anomaly of unusually cold SST in the eastern tropical Pacific (shown in blue, Fig. 2.2e). During the La Niña, the trade wind and Walker cell both intensify, while during El Niño, they reverse direction.

Sir Gilbert Walker (photo: Fig. 2.3) made a pioneering discovery about Walker circulation while he was the director general of India Meteorological Department (IMD), in India (1904–1924). The Walker circulation, which he first described, is named after him. He retired from Kolkata, IMD, in 1924 and joined as a professor in Imperial College London. He also did some ground breaking work on climate variability in the North Atlantic region, which is known as the NAO.

2.1.2 North Atlantic Oscillation (NAO)

Moving from the Pacific region, we are now focusing on the Atlantic Ocean region where another mode of variability, named as the NAO, plays a crucial role. It is a large-scale seesaw in atmospheric mass between the polar low and the subtropical high in the North Atlantic region; it is the dominant mode of climate variability during boreal winter around the North Atlantic ranging from Europe to the central North America and much into northern Asia. It is usually measured as the pressure difference between Azore High and Icelandic Low (regions shown in Fig. 1.1, top).

The major features of the NAO are as follows, also shown in Fig. 2.4: a) the positive phase of NAO shows a stronger than the usual subtropical high-pressure centre over the Atlantic, with a deeper than the normal Icelandic low; b) such increased pressure gradient generates frequent and stronger winter storms that advance the Atlantic Ocean on a more northerly track; c) it causes dry and cold winters in

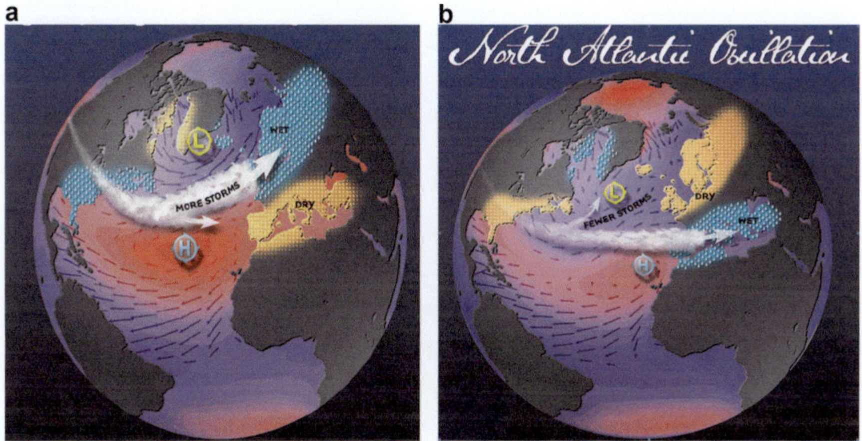

Fig. 2.4 Different phases of the NAO. (**a**) Positive phase. (**b**) Negative phase. (Source: http://www.ldeo.columbia.edu/NAO/, link accessed on 10/12/2017, picture by Visbeck, M)

Greenland and the northern Canada with wet and warm winters in Europe; d) the eastern part of America experiences a humid and mild winter.

Apart from the NAO, which is regionalized and defined only around the Atlantic Ocean region, there is another mode of variability in the NH known as the Arctic Oscillation (AO) that covers more of the Arctic including both the Pacific and the Atlantic Ocean and zonally symmetric, which is described below.

2.1.3   Arctic Oscillation (AO) and Antarctic Oscillation (AAO)

A barometric seesaw between the midlatitudes and polar region is seen in both hemispheres: for the northern hemisphere, it is termed the Arctic Oscillation (AO), whereas in the southern hemisphere, it is the Antarctic Oscillation (AAO). The AO is usually defined as the first leading mode from the EOF analysis of monthly mean geopotential height anomalies at 1000 hPa, poleward of 20° N, in the NH, whereas for the AAO, it is the same at 700 hPa level and measured poleward of 20° S in the SH. In SH, the different pressure level is used to alleviate partially the ambiguities introduced by the reduction of sea level over the high terrain of Antarctica. Both of their positive phases show deeper than normal polar low with stronger than usual midlatitude high. The surface signature of AO and AAO measured in terms of geopotential height, as observed by Thompson and Wallace (2000), is shown in Fig. 2.5.

The AAO and AO patterns are similar from the Earth's surface up to 50 km and in a broader sense are known as the Southern Annular Mode (SAM) and Northern Annular Mode (NAM), respectively. In the stratosphere, the NAM (SAM) is a measure of the strength of the polar vortex, whereas at the surface, the NAM (SAM) is known as the AO (AAO).

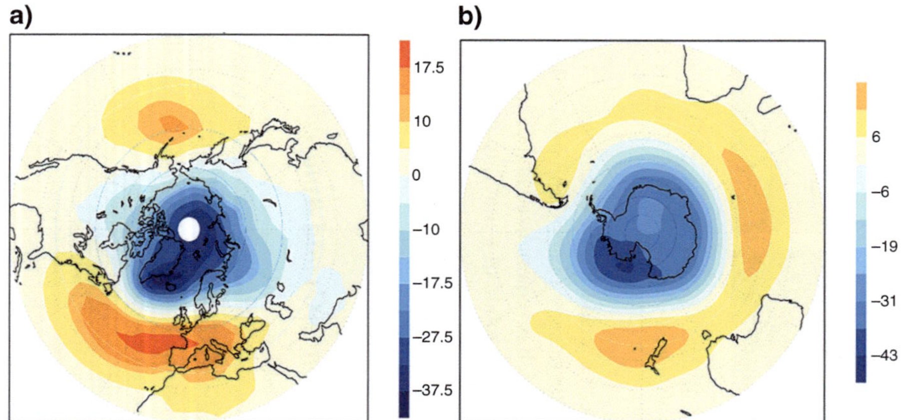

Fig. 2.5 The surface signature of positive AO (**a**) and AAO (**b**) measured in terms of geopotential height. (Source: http://research.jisao.washington.edu/wallace/ncar_notes/, link accessed on 10/12/2017, credit J. Wallace, University of Washington)

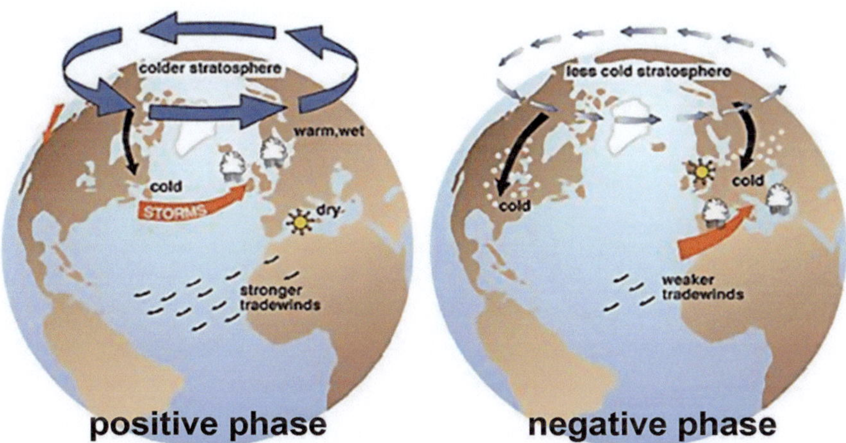

Fig. 2.6 Effects of different phases of the Arctic oscillation. (Source: http://nsidc.org/arcticmet/patterns/arctic_oscillation.html, link accessed on 10/12/2017, Image/photo courtesy of J. Wallace, University of Washington, supplied by the National Snow and Ice Data Center, University of Colorado, Boulder)

The band of upper-level winds that circulate the pole in the stratosphere generate the polar vortex (also shown in Fig. 2.6 with blue arrows). When the annular mode index becomes positive, the strength of vortex increases and winds constrict around the pole, locking cold air masses in places near the pole. On the other hand, a negative surface annular mode, associated with weak vortex, allows intrusion of cold air masses to plunge southward into North America, Europe and Asia (for NAM). Different phases of annular modes are thus linked with variations of surface weather patterns in the polar region. The associated surface climate change with the phase of annular modes is described here with an illustration in Fig. 2.6. In the positive phase of the surface NAM, the higher pressure at the midlatitudes drives cyclones farther north towards the Arctic that alters the pattern of circulation. It then brings wetter conditions to the Scandinavia, Iceland, and Alaska; alongside it brings drier weather to the Mediterranean and the Western USA, while the situation changes during the negative phase (Fig. 2.6). Moreover, in the positive phase of oscillation, cold air masses during winter do not enter as far into the Europe and North America as it would during the negative phase. It preserves much of Europe and the eastern Rocky Mountains of the USA warmer than normal during periods of high AO. Alongside, it keeps places like Newfoundland, Greenland and Labrador colder than usual.

2.1.4 Pacific Decadal Oscillation (PDO)

Apart from the ENSO, there is another mode of strong variability in the Pacific, known as the PDO. The PDO is an ENSO-like pattern of the Pacific climate variability which is long-lived and is the leading principal component of the North Pacific monthly SST variability from poleward of 20°N.

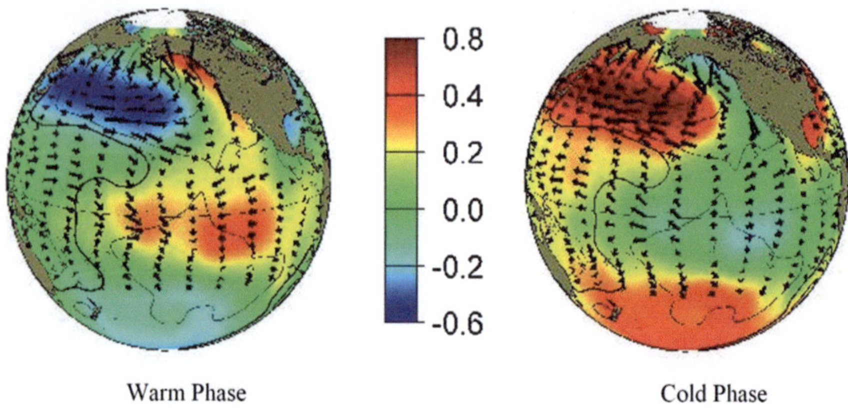

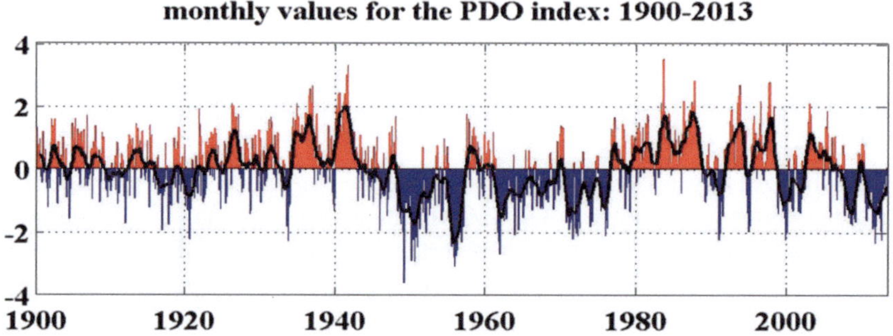

Fig. 2.7 Typical wintertime SST in °C (colours), surface wind stress (arrows) anomaly and SLP (contours) during warm and cold phases of the PDO (Top). Time series of the PDO are shown in bottom. (Source: http://jisao.washington.edu/pdo/, link accessed on 10/12/2017, credit: Nate Mantua, NOAA)

Major features of the PDO in relation to the ENSO may be described as follows: a) its main climatic fingerprints are in the north Pacific with secondary signatures (of SST) in the tropics (shown in Fig. 2.7), which is the other way round for the ENSO (Fig. 2.2d, e), though both their warm and cold phases possess similar sign temperature anomaly like ENSO (and hence the phases have been named); b) it has a long-term variability and persists for 20–30 years, whereas for ENSO the variability is for 2–7 years.

2.1.5   Atlantic Multidecadal Oscillation (AMO)

The AMO is defined as the area average over the entire North Atlantic of the low-pass filtered annual mean SST anomalies after removing linear trend (Fig. 2.8). The time scale of the AMO is ~ 20–40 years.

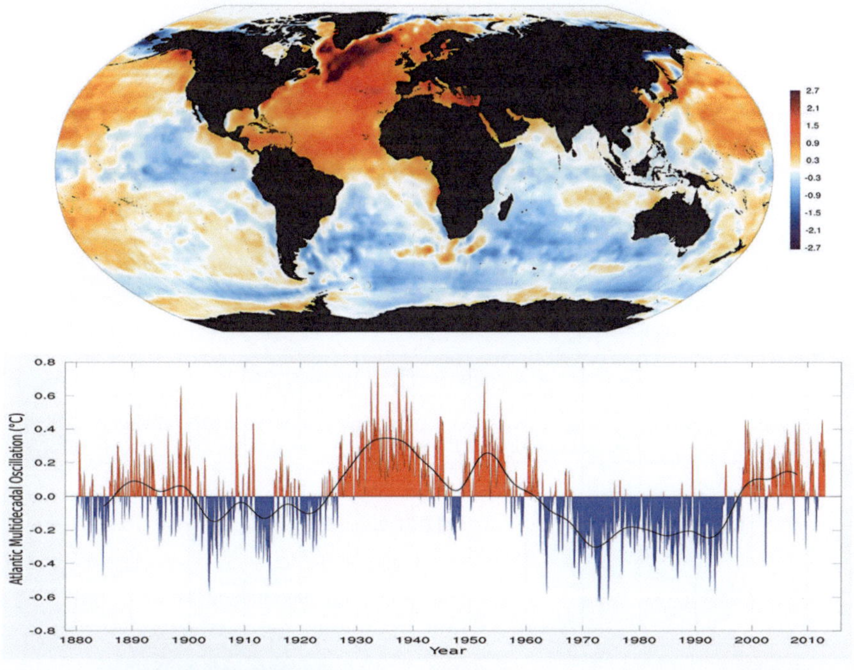

Fig. 2.8 The spatial pattern of the AMO shown at the top and temporal behaviour at the bottom. (Source: https://en.wikipedia.org/wiki/Atlantic_multidecadal_oscillation, link accessed on 10/12/2017, credit: Giorgiogp2 under CC BY-SA 3.0)

Apart from that major variability, as discussed above, here are descriptions of two more important climate variability known as the Indian Summer Monsoon and the Indian Ocean Dipole.

2.1.6   Indian Summer Monsoon (ISM)

Monsoon means seasonal wind reversals, and for ISM, it is from north easterly (NE-ly) during northern winter to south westerly (SW-ly) during northern summer as shown in Fig. 2.9. It is associated with the movement of intertropical convergence zone (ITCZ) which separates northern hemisphere to that from the southern hemisphere. Indian subcontinent experiences heavy rainfall during summer (June–July–Aug–Sept) as moisture-rich air from ocean enters the land region. The Walker circulation and Hadley circulation both play a role in controlling ISM rainfall variability.

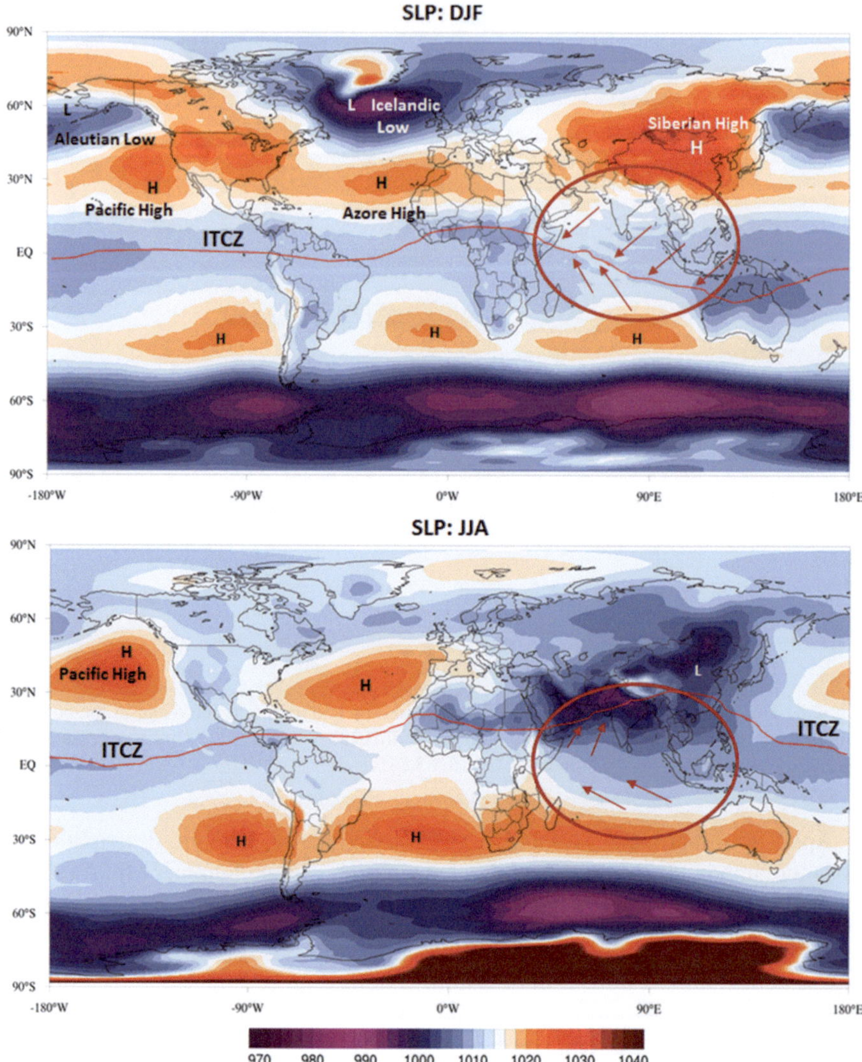

Fig. 2.9 The climatology of SLP during northern winter (top) and northern summer (bottom) is shown again (as in Fig. 1.1) with particular emphasis on ISM region. A shift in the ITCZ and associated seasonal wind reversal is marked by red oval, covering Indian subcontinent. (Source: Background maps generated via http://cci-reanalyzer.org/, using NCEP/NCAR Reanalyses v1 data, ClimateReanalyzer.org, University of Maine, Climate Change Institute)

2.1.7   Indian Ocean Dipole (IOD)

The IOD is a dipole behaviour in the Indian Ocean that regulates rainfall around Australia, East Africa and Indian subcontinent as shown in Fig. 2.10.

During the positive (negative) phase of the IOD, Darwin in Australia experiences drought (heavy rain) which are also associated with cold (warm) ocean water. On the other hand, in India or East Africa, there is heavy rainfall (drought).

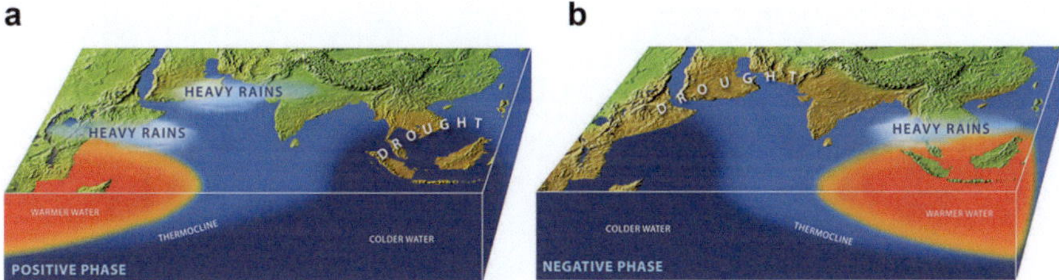

Fig. 2.10 Different phases of the Indian Ocean Dipole. (**a**) Positive phase. (**b**) Negative phase. Copyright © Woods Hole Oceanographic Institution. (Source: http://www.whoi.edu/, uploaded on 10/12/2017, credit E. Paul Oberlander, Woods Hole Oceanographic Institution. Copyright © Woods Hole Oceanographic Institution)

2.2 Variability in the Stratosphere

While discussing the polar modes of variability, it is described that the troposphere and stratosphere are strongly coupled. Thus for improving understanding of tropospheric variability, it is important to know about the variability in the stratosphere. Various tropospheric and stratospheric variabilities are also discussed in Roy (2010). There are two primary forms of variability in the stratosphere, viz. the **Q**uasi-**B**iennial **O**scillation (**QBO**) and the **S**tratospheric **S**udden **W**arming (**SSW**), and they are discussed here briefly.

2.2.1 Quasi-Biennial Oscillation (QBO)

It is an oscillation in the equatorialstratosphericwind, with zonal winds change between the east and west with a period of just greater than 2 years. The time series of equatorial deseasonalized zonal wind is shown in Fig. 2.11, where the colours in the lower and middle stratosphere reveal the two different phases of the QBO: red for westerly and blue for easterly.

The main characteristics of the QBO are as follows:

(a) It is prominent between 10 mb and 100 mb.
(b) The period of oscillation is around 20–36 months with a mean of around 28.
(c) Wind regimes propagate downward with time with speed roughly 1 km/ month.
(d) The maximum peak-to-peak amplitude of 40 to 50 m/s is seen at 20mb.
(e) Easterlies are usually stronger than westerlies.
(f) Easterly winds last longer at lower levels than westerlies, while the reverse is true at higher levels.
(g) QBO shows considerable variability, in both amplitude and period.

Mechanisms for QBO Formation
The QBO is mainly governed by equatorially trapped Kelvin and Rossby-gravity waves; the former provides the westerly momentum, the latter the easterly.

In Fig. 2.12, the formation of QBO is shown in a series of step (from a to f) and is described below. Black wavy line shows mean flow, with easterly (or westerly)

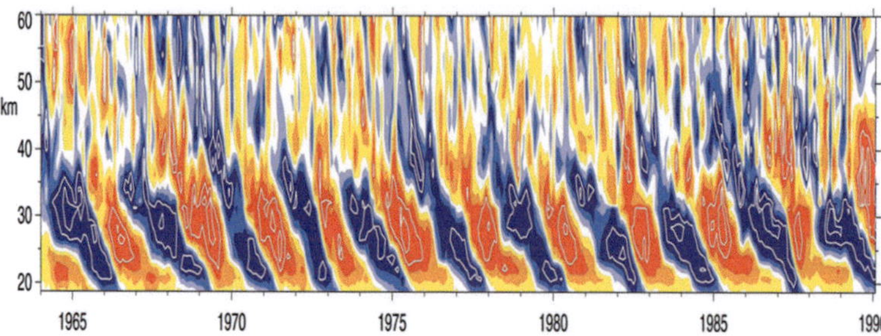

Fig. 2.11 Time-height section of the monthly mean zonal wind components (m s^{-1}), with the seasonal cycle, removed, for 1964–1990. The contour interval is 6 m s^{-1}, with the band between −3 and +3 unshaded. (Source: Baldwin et al. 2001 and Gray et al. 2001)

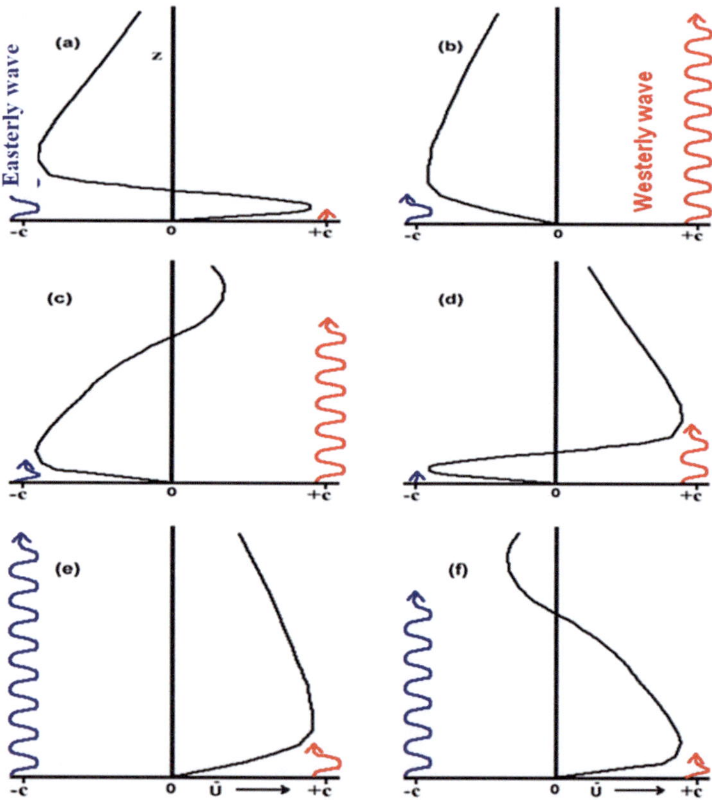

Fig. 2.12 Theory of QBO formation in a series of steps (a to f). (Source: http://ugamp.nerc.ac.uk/hot/ajh/qbo.htm, uploaded on 10/12/2017, after Plumb (1984))

flow shown along the negative (or positive) side of the abscissa. The equatorially trapped easterly wave, propagating upward, is marked by blue, whereas the same for westerly is of red. Both the westerly and easterly maxima of mean flow are descending (a). Upward travelling waves are seen depositing their momentum just below the maxima. Viscous diffusion destroys the westerlies, when the shear zone of the mean westerly flow is sufficiently narrow. Upward propagating westerly waves can travel to high levels (b) through the mean easterly flow. The more freely upward travelling westerlies lead to a new westerly regime of mean flow, as those dissipate at higher altitudes and produce a westerly acceleration (c). (d) shows both systems of the mean flow descending downwards until the easterly shear zone of the mean flow becomes sufficiently narrow to destroy easterlies and upward propagating easterlies can then spread to high altitudes through westerly mean flow (e). Finally, it leads towards the formation of a new easterly regime of the mean flow shown in (f); likewise, the same process (as described in (a) to (f)) continues.

The QBO is directly measured in operational wind measurements by radiosondes at equatorial meteorological observatories. Observations are taken within 2-degree latitude from the equator. A stratospheric research group at the Free University of Berlin, Barabara Naujokat, has processed and collected radiosonde measurements from 1953 onward. It is from Kanton Island, Gan (Maldives) and Singapore (Naujokat 1986; Labitzke and collaborators 2002). The respective locations and available periods for these three observing stations are 171°43'W/ 02°46'S, Jan.1953–Aug.1967; 73°09'E/ 00°41'S, Sept.1967–Dec.1975; and 103°55' E/ 01°22' N, Jan.1976–Dec.2004. The QBO suggests a high degree of zonal symmetry which allows the merger of the equatorial zonal wind profiles of these three individual stations into one dataset covering a longer period. The data of QBO since 1953 at various levels from 10 hPa to 70 hPa is available from http://www.pa.op.dlr. de/CCMVal/Forcings/qbo_data_ccmval/u_profile_195301-200412.html.

Reconstructions of the QBO dating back to 1900 are now available (Brönnimann et al. 2007). It is based on the data from historical pilot balloon and the data of hourly SLP from Jakarta, Indonesia. The latter was considered to extract the solar semi-diurnal tide-related signal. This is because, in the middle atmosphere, it is modulated by the QBO. There is a good match with these reconstructions with the extracted QBO signal from historical total ozone data dating back to 1924. Moreover, the maximum phases of the QBO are also a well captured after about 1910.

Wind and Temperature (Vertical Structure)
Before discussing SSW, it is better to have some knowledge about the region of the polar vortex and the vertical distribution of wind and temperature profile as presented in Fig. 2.13. In the troposphere, the temperature decreases with height up to tropopause region. This region is a boundary between the troposphere and stratosphere. It is 18 km around the tropics, through 8 km in the polar region. Due to an abundance of ozone in the stratosphere, the temperature increases there with height.

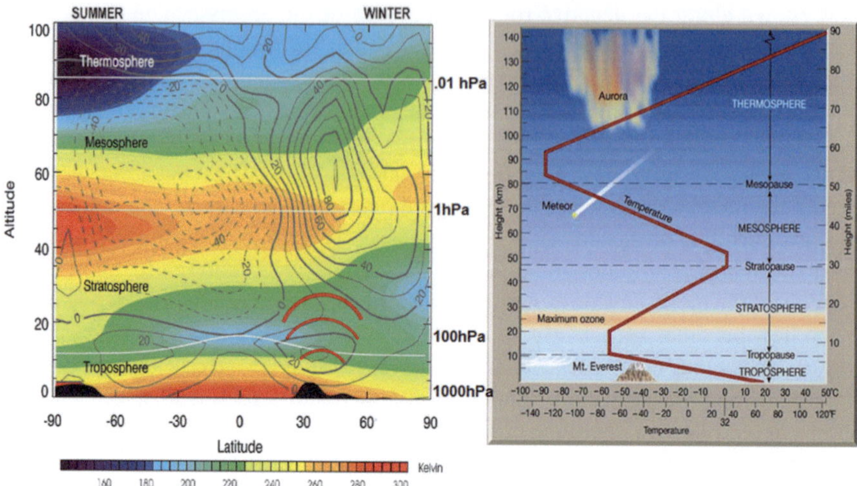

Fig. 2.13 The wind and temperature distribution in the top of atmosphere. (Source: Left: http://lasp.colorado.edu/sorce/news/2004ScienceMeeting/SORCE%20WORKSHOP%202004/SESSION_2/2_4_McCormack.pdf Credit, John McCormack, US Naval Research Laboratory, Washington DC.; Right: http://www.4college.co.uk/as/atm/air.php, Link accessed on 10/12/2017)

It reaches maximum in thestratopausearea (50 km and 1 hPa level). It then decreases with height in the mesosphere, which is up to 80 km. Above 80 km, which is the thermosphere, temperature again increases with height. The polar region around the stratopause, where polar vortex is formed in both the hemispheres, is responsible for SSW as discussed below.

2.2.2   Stratospheric Sudden Warming (SSW)

It is the most dramatic event in the stratosphere during winter hemisphere, where the polar vortex of westerly winds, over the course of a few days, abruptly slows down or even reverses direction. It is also associated with a rise in the stratospheric temperature by several tens of Kelvins. Figure 2.14 illustrates the latitude time series of average zonal temperatures around 30 km during two different periods. The periods of sudden warming are marked by 'X' and 'Y' in Fig. 2.14a, b, respectively, when temperatures were observed to rise abruptly.

SSWs can be classified into three broad categories:

- *Major Warming*

The major warming occurs when westerly winds at 60°N and 10 hPa reverse, i.e. become easterly from westerly. A thorough disruption of the polar vortex is noticed and the vortex either splits into two separate vortices or displaces from its normal location over the pole.

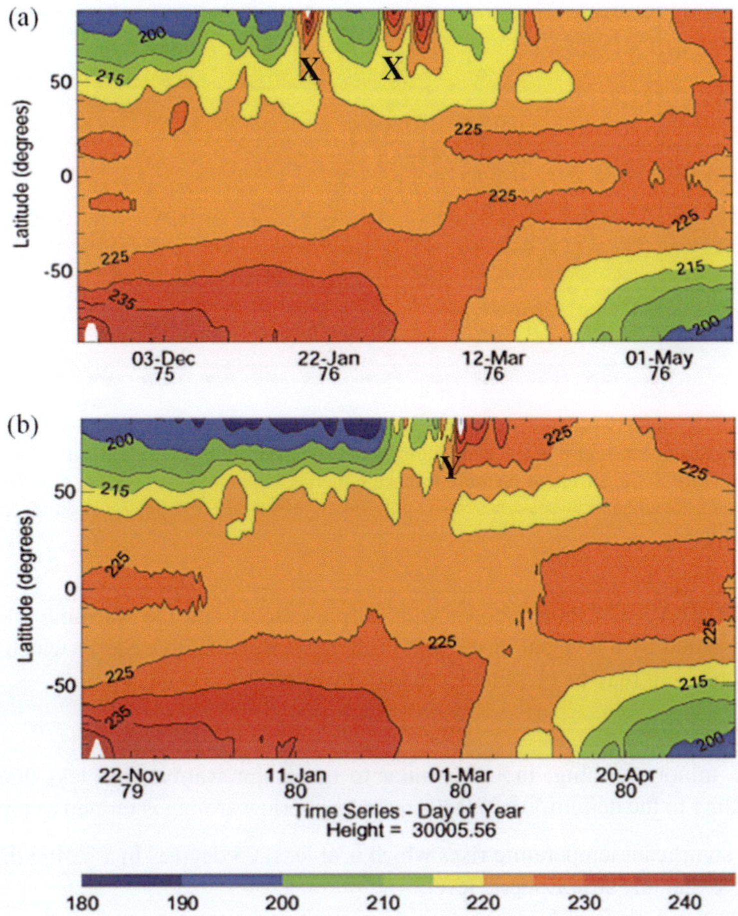

Fig. 2.14 Latitude-time series of average zonal temperatures around 30 km. (Source: http://lasp.colorado.edu/sorce/news/2004ScienceMeeting/SORCE%20WORKSHOP%202004/SESSION_2/2_5_Gray.pdf, link accessed on 10/12/2017, credit: Lesley Gray)

For a major warming to occur, the following two defining criteria need to be satisfied as specified by WMO:

(a) An increase in temperature, poleward from 60° latitude at 10 hPa of 40–60 K, takes place in less than 1 week.
(b) Zonal mean zonal wind over the same region reverses.

Such warming can be shown in Fig. 2.15, where the time series of daily temperature (°C) over the north poles is plotted from November 2005 till April 2006. The variations at different levels, viz. 1 hPa, 10 hPa and 30 hPa, are shown by different colours. A grey line marks the mean of 30 years north polar temperature at 30 hPa level. From this picture, it is clear that there is a sudden rise in temperature for a few weeks after 1 Jan. 2006 at all three pressure levels. Since the temperature rose more

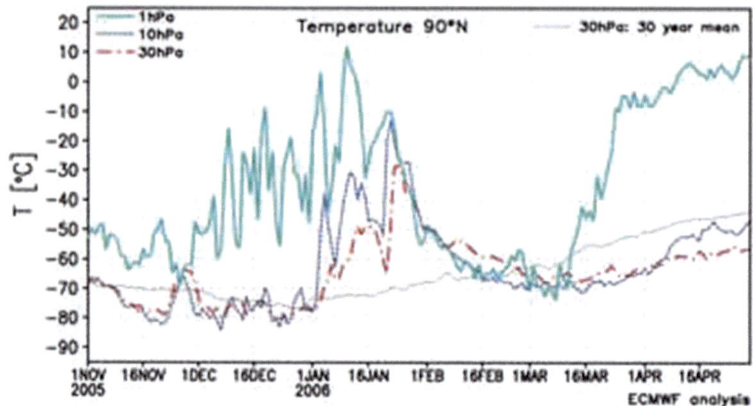

Fig. 2.15 Time series plot of the daily temperatures (°C) over the North Pole during November 2005 till April 2006. Different pressure levels are marked with colours. (Source: http://www.bu. edu/cawses/documents/cawses-news-v3-n2.pdf, link accessed on 10/12/2017, Credit: K. Labitzke and M. Kunze)

than 40° at 10 hPa level (shown with purple colour) and the warming was also accompanied by a reversal of the wind (i.e. from westerly to easterly which is not shown here), it may be classified as a major warming.

- *Minor warming*

The minor warming, though similar to the major warming, is less dramatic. According to the definition of WMO, a stratospheric warming is termed as minor if:

(a) A significant temperature rises which is at least 25 degrees in a period of week or less at any stratospheric level.
(b) The wind reversal from westerly to easterly is less extensive (i.e. the zonally averaged zonal wind does not reverse) and the polar vortex is not broken down.

- *Final warming*

The radiative cycle in the stratosphere means that the mean flow is westerly during the winter, while easterly during the summer. A final warming takes place on this transition, so that winds around the polar vortex reverse direction for the warming and the stratosphere enters the summer easterly phase. It is known as the final warming of the current winter, because winds do not change back until the following winter (shown by 'Y' in Fig. 2.14b).

Referenes

Baldwin MP et al (2001) The quasi-biennial oscillation. Rev Geophys 39(2):179–229
Brönnimann S, Annis JL, Vogler C, Jones PD (2007) Reconstructing the Quasi-Biennial Oscillation back to the early 1900s. Geophys Res Lett 34:L22805. https://doi.org/10.1029/2007GL031354

Gray LJ et al (2001) A data study of the influence of the equatorial upper stratosphere on northern hemisphere stratosphereic sudden warmings. QJR Meteorol Soc 127:1985–2003. https://doi.org/10.1002/qj.49712757607

Labitzke K and collaborators (2002) The Berlin stratospheric data series. Data available from Meteorolog. Institute, Free University Berlin, http://strat-www.met.fu-berlin.de/labitzke/special/JASTP-SPECIAL-Labitzke-2004.pdf

Naujokat B (1986) An update of the observed quasi-biennial oscillation of the stratospheric winds over the tropics. J Atmos Sci 43:1873–1877

Roy I (2010) Solar signals in sea level pressure and sea surface temperature. Department of Space and Atmospheric Science, PhD Thesis, Imperial College, London

Thompson DWJ, Wallace JM (2000) Annular modes in the extratropical circulation. Part I: month-to-month variability. J Clim 13:1000–1016

Chapter 3
Stratosphere-Troposphere Coupling

Abstract This chapter discusses stratosphere-troposphere coupling starting from a very basic background knowledge. Later it focused on the individual role of the Sun and QBO to regulate such coupling. It explores Holton-Tan effect. The combined influence of the Sun and QBO to regulate upper stratospheric polar temperature was also attended. The role of polar annular modes was described that communicate such signal down to the troposphere. Various other routes are also discussed to show how upper stratospheric solar signature from the polar vortex could be transported downward and affect tropospheric climate.

Keywords Holton-Tan effect · Charney-Drazen criteria · Planetary wave · QBO · annular mode · Polar vortex · Zero wind line · Rossby wave · Brewer-Dobson circulation

3.1 Background

The very basics of the planetary wave (PW) propagation, mainly governed by the Charney Drazin criteria, are shown here by the flow chart of Fig. 3.1. It states that the upward movement of the stationary waves needs to satisfy two main criteria: (1) the background flow must be westerly, though not too strong, and (2) only long waves, i.e. wave number 1–3, can propagate. Now, winter NH satisfies both the criteria as mentioned above: there are westerly in the high and midlatitude stratosphere; moreover around the midlatitude, the orography together with land-sea temperature difference generates long Rossby waves. Thus during the winter NH, there lies the proper breeding ground for the upward propagation of the stationary waves – that allow strong coupling from the troposphere to the stratosphere be possible.

© Springer International Publishing AG, part of Springer Nature 2018

I. Roy, *Climate Variability and Sunspot Activity*, Springer Atmospheric Sciences, https://doi.org/10.1007/978-3-319-77107-6_3

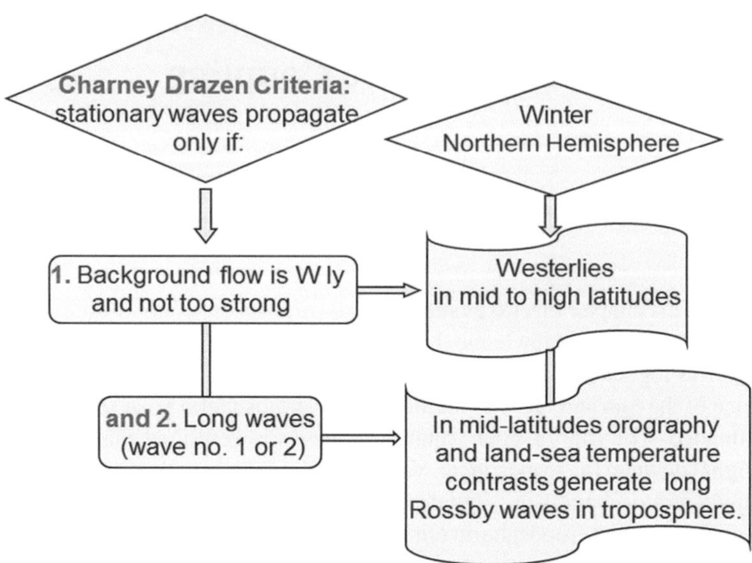

Fig. 3.1 Flow chart showing the background of planetary wave propagation (Roy 2010)

3.2   Discussion with Schematic

Here in schematic Fig. 3.2, we illustrate the coupling mechanism of the middle atmosphere. A picture with latitude vs. altitude (in km) is presented where different regions, viz. troposphere, the stratosphere, mesosphere and thermosphere, are marked based on their height levels. The zonally averaged (2D) zonal wind (ms^{-1}) for December is shown by solid lines (westerly) or dashed lines (easterly) with respective magnitudes. Zero wind lines that separate westerly from the easterly have been marked with label zero. Zonally averaged (2D) zonal temperature (K) is shown with appropriate colours, chosen from the colour bar, placed at the bottom of the picture. From the description, it is seen that there is an ascending motion in the thermosphere during the summer hemisphere, while descending during the winter hemisphere, shown with thick arrows (however, in our subsequent analysis, we are not attending the thermosphere). But the Brewer-Dobson circulation (shown with narrower arrows in the schematic), the primary circulation in the stratosphere, is important in our analysis. As observed in the picture, it is a loop-like circulation originating from the equatorial region of the lower stratosphere, travels towards the polar stratosphere (up to stratopause), and then finally comes back in the lower stratosphere.

In Fig. 3.1, we described how the NH winter can be susceptible to the PW propagation, and the upward propagation of these PWs is shown here in Fig. 3.2 with red (around 30°N). These PWs move upward and finally break around places of the polar stratospheric jet, decelerating westerlies around that region, alongside warming. Now, PW propagation is sensitive to two major influences; one is solar, and the other is the QBO, which is discussed below.

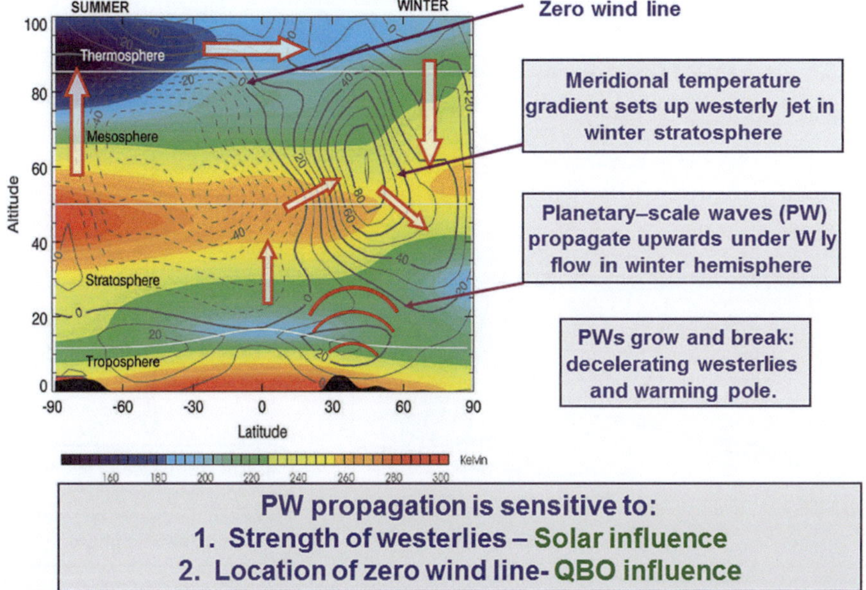

Fig. 3.2 Schematic representation of middle atmosphere coupling. (Source: http://lasp.colorado.edu/sorce/news/2004ScienceMeeting/SORCE%20WORKSHOP%202004/SESSION_2/2_4_McCormack.pdf, picture (left); also after, CIRA climatology, Flemming et al., 1990)

3.3   Strength of Westerly: Solar Influence

The solar UV at 205 nm increases ~6% from solar min to solar max, which causes more ozone heating in the upper stratosphere (shown with red colour in summer hemisphere around stratopause region in Fig. 3.2). That indicates more warming around the equatorial upper stratospheric region during higher solar years compared to the low solar years. Such a strengthening in the latitudinal temperature gradient, around the polar stratosphere, is liable to alter the wind structure, strengthening the winter stratospheric polar jet. Such intensified polar stratospheric jet in higher solar years can be responsible for interaction and breaking of PW; subsequently causing warming around winter upper stratospheric pole. Though the Sun through interaction with PW can play a crucial role in regulating polar temperature, it is not the sole factor responsible for perturbing polar stratosphere. PW propagation is also susceptible to the phase of QBO, and the relevant mechanism involved is described below:

3.4   Role of Zero Wind Line: QBO Influence

The QBO has an influence on PW propagation, via modulating the zero wind line. It is mainly governed by well-known 'Holton-Tan effect' (Holton and Tan 1980, 1982) which states QBO E-ly is associated with the warm polar stratosphere.

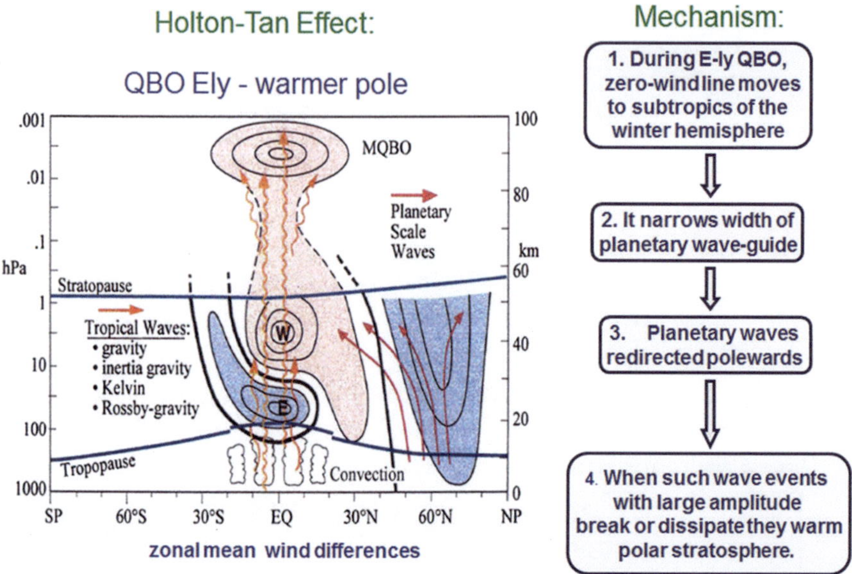

Fig. 3.3 Zonal mean wind differences and illustration of Holton-Tan effect. (Source *(left)*: Baldwin et al. 2001)

Schematically, it is shown in Fig. 3.3, with a mention about the major steps involved. It is also accompanied by a picture from Baldwin et al. (2001), which is a plot of zonal mean wind differences in a latitude vs. height/pressure cross section. Here easterly wind is marked by blue; PWs are shown by red, and various tropical waves originated in the equatorial troposphere (not relevant in describing Holton-Tan effect) are shown by yellow.

During the easterly phase of QBO, the zero wind line (that separates easterly from the westerly) moves towards subtropics. It narrows the width of PW guide. Thus, PWs are more redirected towards the pole during the E-ly phase of QBO compared to the W-ly phase. When such wave events with large amplitude dissipate or break, they warm the polar stratosphere, depositing their easterly momentum around the polar region (shown with blue colour there). This is the mechanism of the so-called Halton-Tan effect – i.e. warm pole during easterly QBO.

We described the primary mechanisms relating to the interaction of the QBO and Sun, with upward propagating PW – which was shown to be responsible for modulating the polar stratospheric temperature during the winter hemisphere.

3.5   Sun, QBO and Polar Temperature in North Pole

Labitzke and van Loon (1992) plotted a scatter diagram (shown in Fig. 3.4, top) that depicts values of 30 hPa temperature at the North Pole for each year in Jan/Feb as ordinate and solar 10.7 cm flux along the abscissa. The phase of the QBO was

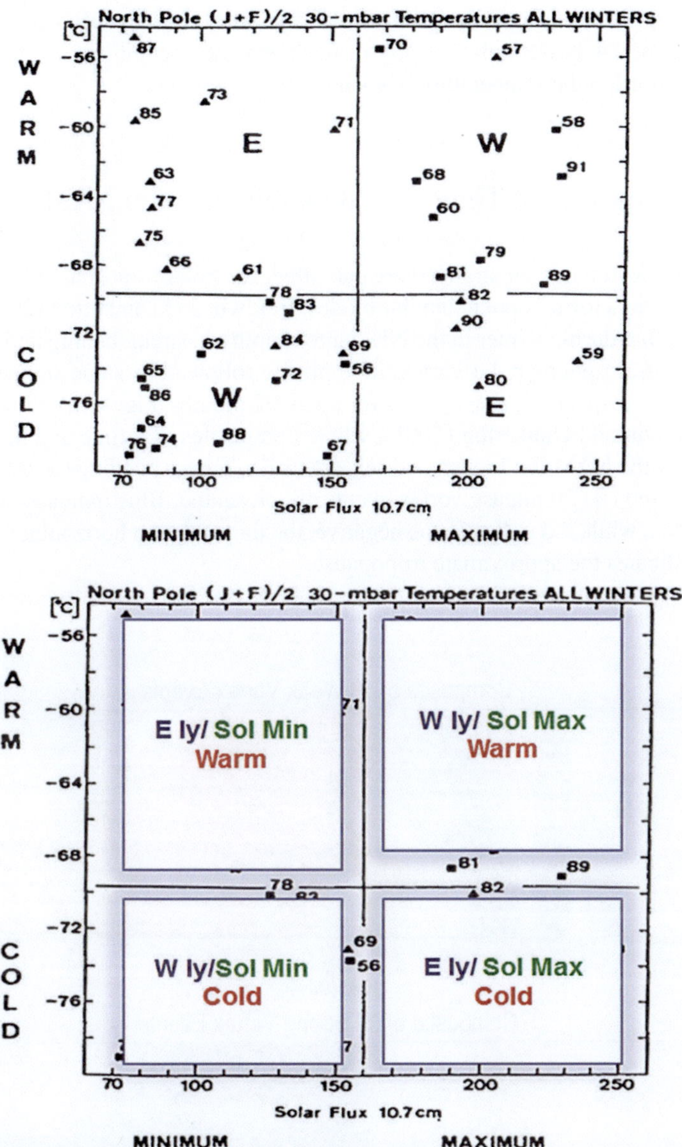

Fig. 3.4 Solar-QBO relationship in the North Pole following Labitzke and van Loon (1992) (top); same plot, but mentioning various combinations in each quarter (bottom)

marked with symbols; square for westerly and triangle for easterlies. The demarcating vertical and horizontal line have been shown to mark regions where certain phases of the QBO predominate and shown by the E and W labels. From Fig. 3.4, it is clear that warm polar temperatures tend to occur during the westerly QBO phase at solar maximum and east QBO phase at solar minimum. Whereas the cold polar temperature occur during W-ly QBO at solar minimum and E-ly QBO at solar maximum. Figure 3.4, bottom plot, clearly designates various combinations of the Sun, QBO and north polar temperature in each quarter of top plots.

3.6 Composites of Time Height Development of NAM

Perturbations in the polar stratosphere can affect the polar troposphere for the next few months in a form of polar annular modes. Baldwin and Dunkerton (2001, 2005) suggested that during winter in the NH, large-amplitude strengthening and weakening of the stratospheric polar vortex are typically followed by same signed anomalies in the troposphere. It usually lasts for up to 3/2 months. Figure 3.5 shows results from Baldwin and Dunkerton (2001), where composites of a time and height progression of the NAM for 18 weak vortex events (A) (corresponding to stratospheric warming) and (B) 30 intense vortex events are presented. Blue indicates a positive NAM index, while red indicates the negative; the thin and faint horizontal line above 10 km indicates the approximate tropopause.

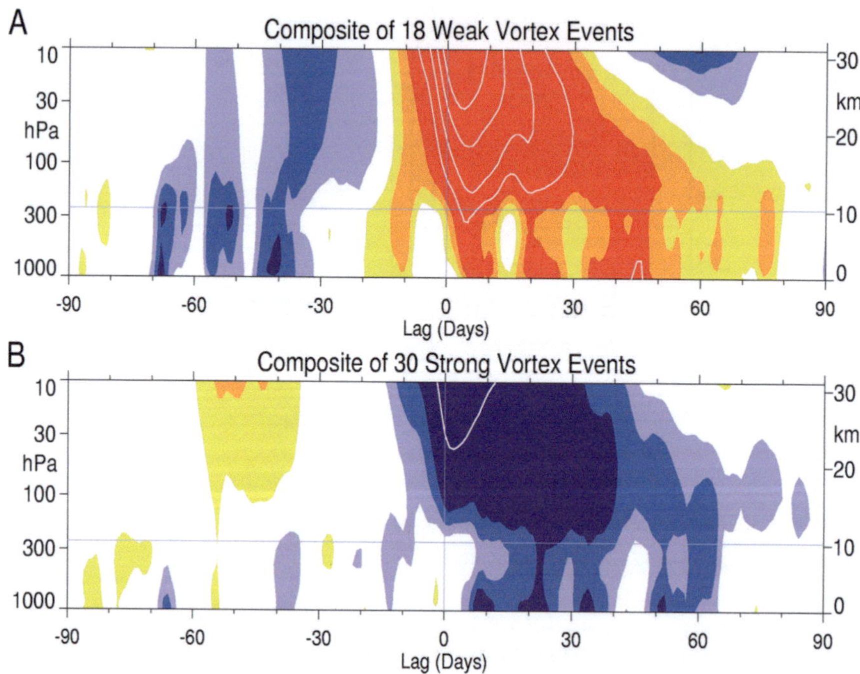

Fig. 3.5 Composites of time and height growth of NAM for 18 weak vortex events (**a**) and for 30 strong vortex events (**b**). (Source: From Baldwin and Dunkerton 2001. Reprinted with permission from AAAS)

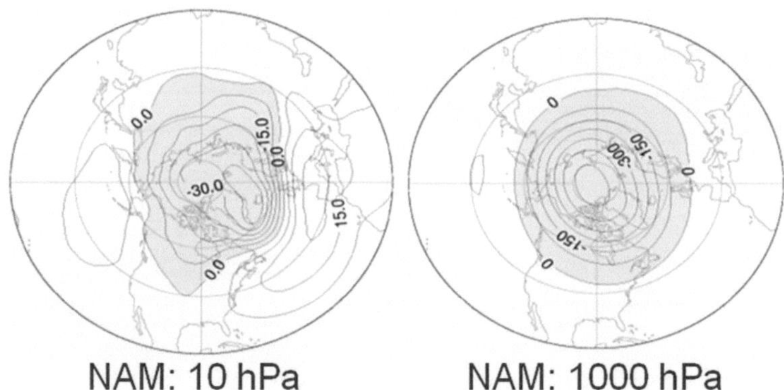

Fig. 3.6 NAM at 10 hPa and 1000 hPa for November–April, 1958–2000 (with permission from Elsevier: Baldwin and Dunkerton 2005)

Such observations suggest that large-amplitude variations in the stratospheric polar vortex are followed by persistent anomalies not only in the stratosphere but also in the troposphere. In Fig. 3.5, the onset of a stratospheric event at 10 hPa corresponds to day 0. On an average, it is ~2–3-month period following the start of the composite stratospheric event that can be linked with similarly signed tropospheric anomalies.

3.7   Annular Modes Pattern Similar

The study of Baldwin and Dunkerton (2005) also shown here in Fig. 3.6 depicts that the NAM patterns measured as the leading empirical orthogonal function of Nov–April (low-pass filtered, 90 days) geopotential at 10 and 1000 hPa are similar, and to first order, zonally symmetric.

3.8   Solar Influence: Polar Vortex and Tropical Lower Stratosphere

The schematic of Fig. 3.7 shows a flow chart that depicts a comprehensive overview of proposed mechanism by Kodera and Kuroda (2002) relating to the polar stratosphere and, finally, warming of the equatorial lower stratosphere by solar forcing. The relevant picture from Kodera and Kuroda (2002) is also presented in that schematic. It is a latitude vs. height plot, where the summer hemisphere and the winter hemisphere are also marked. As solar UV increases from solar min to solar max causing more ozone heating in the upper stratosphere during higher solar years – it is shown with red colour in the summer hemisphere. Such warming can alter the

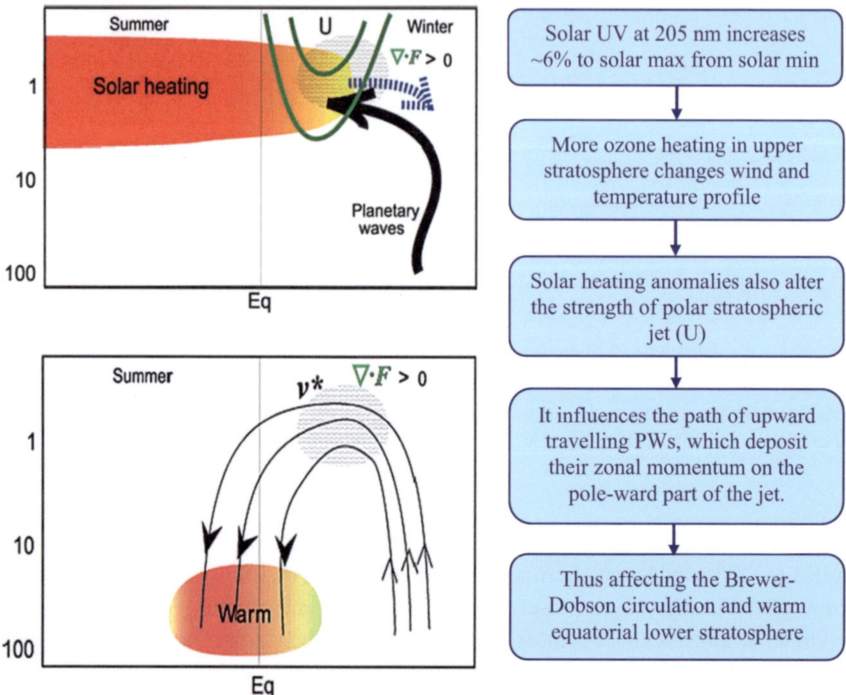

Fig. 3.7 Flow chart showing proposed mechanism by Kodera and Kuroda (2002) relating to warming of the equatorial lower stratosphere by solar forcing. (Picture (left): Kodera and Kuroda 2002)

temperature and wind structure along with changes in the strength of polar stratospheric jet (shown with U). It, subsequently, influences the route of the upward travelling planetary waves (shown with thick black lines) which deposit their zonal momentum on the poleward part of the jet and thus reducing the strength of the Brewer-Dobson circulation. Such reduction in the strength of the Brewer-Dobson circulation, for HS years, can be responsible for the warming of the equatorial region of the lower stratosphere. Finally, such warming can be transported to the troposphere via circulations, i.e. the Hadley cell and the Ferrel cell, as elaborately studied by Haigh and co-workers in a succession of papers (1996, 1999, 2005).

3.9   Solar Influence: Tropical Lower Stratosphere to Troposphere

Modelling Study: Haigh (1996, 1999), Simpson et al. (2009)

Haigh (1999) shows results from atmospheric circulation model for the influence of the 11-year solar cycle on the climate of the lower atmosphere. A pattern of response is noticed in which the tropical Hadley cells broaden and weaken for high irradiance. It also revealed that the midlatitude Ferrel cells and subtropical jets move poleward during active periods. Such a feature, as also noticed in the modelling work by Haigh (1996), is very similar to the observational study. Without any polar vortex and with fix SST temperature, they only imposed heating in the tropical lower stratosphere in their model. The changes in dynamics are responsible for subtropical warming and a characteristic vertical band structure of midlatitude temperature variations. Solar forcing in the lower stratosphere can modify the strength, and the location of Hadley and Ferrel cell along with the subtropical jet has also been identified in the recent modelling study of Simpson et al. (2009).

Observational Study: Haigh (2005), Bronnimann et al. (2007)

For NCEP reanalysis data of zonal mean zonal winds, using multiple regression analysis, Haigh (2005) showed when the Sun is more active, the midlatitude jets are positioned further poleward and become weakened and deduce an expansion in Hadley cells and a polar shift in the Ferrel cells. Brönnimann et al. (2007) using upper air data also noticed poleward displacement of the subtropical jet and Ferrel cell with increasing solar irradiance.

References

Baldwin MP, Dunkerton TJ (2001) Stratospheric harbingers of anomalous weather regimes. Science 294(5542):581–584. https://doi.org/10.1126/science.1063315

Baldwin MP, Dunkerton TJ (2005) The solar cycle and stratosphere-troposphere dynamical coupling. J Atmos Sol-Terr Phys 67:71–82. https://doi.org/10.1016/j.jastp.2004.07.018

Baldwin MP et al (2001) The quasi-biennial oscillation. Rev Geophys 39(2):179–229

Brönnimann S, Ewen T, Griesser T, Jenne R (2007) Multidecadal signal of solar variability in the upper troposphere during the 20th century. Space Sci Rev 125(1–4):305–317

Haigh JD (1996) The impact of solar variability on climate. Science 272(5264):981–984

Haigh JD (1999) A GCM study of climate change in response to the 11-year solar cycle. Q J Roy Meteorol Soc 125(555):871–892

Haigh JD, Blackburn M, Day R (2005) The response of tropospheric circulation to perturbations in lower-stratospheric temperature. J Climate 18(17):3672–3685

Holton JR, Tan HC (1980) The influence of the equatorial quasi-biennial oscillation on the global circulation at 50 mb. J Atmos Sci 37:2200–2208

Holton JR, Tan HC (1982) The quasi-biennial oscillation in the Northern Hemisphere lower stratosphere. J Meteorol Soc Jpn 60:140–148

Kodera K, Kuroda Y (2002) Dynamical response to the solar cycle. J Geophys Res 107(D24):4749. https://doi.org/10.1029/2002JD002224

Labitzke K, van Loon H (1992) On the association between the QBO and the extratropical stratosphere. J Atmos Terr Phys 54(11/12):1453–1463

Roy I (2010) Solar signals in sea level pressure and sea surface temperature. Department of Space and Atmospheric Science, Ph.D. Thesis, Imperial College London

Simpson IR, Blackburn M, Haigh JD (2009) The role of eddies in driving the tropospheric response to stratospheric heating perturbations. J Atmos Sci 66(5):1347–1365

Chapter 4
Teleconnection Among Various Modes

Abstract This chapter discusses various teleconnections involving climate modes. First, it focuses different teleconnections involving polar vortex, QBO and ENSO; later the discussion was extended up to troposphere. The ENSO being one of the major tropospheric variabilities, the subsequent discussion covered various characteristic features of ENSO and subsequently ENSO-related teleconnections. The topics covered are ENSO and jet, influence of ENSO around the world, ENSO and ISM and various types of ENSO including central Pacific (CP) type (or Modoki ENSO) and east Pacific (EP) type (or canonical ENSO). Specific influences of these two types of ENSO are also attended in details.

Keywords El Niño · La Nino · Canonical ENSO · Modoki ENSO · teleconnection · polar vortex · Indian Summer Monsoon

4.1 Polar Vortex, QBO and ENSO

Thompson et al. (2002) noted that pronounced weakening of the NH stratospheric polar vortex in winter tends to be followed by episodes of anomalously low surface air temperatures throughout densely populated regions such as the Northern Europe, eastern North America and East Asia that persists for ~2 months. Strengthening of the vortex tends to be followed by anomalous surface temperature in the reverse way. The QBO in the equatorial stratosphere during midwinter has a similar but somewhat lesser influence, presumably through its impact on the stability and strength of the stratospheric polar vortex. It suggests that the easterly QBO phase favours an increased occurrence of extreme cold events and vice versa for westerly phase. The signature of the QBO in wintertime NH temperatures is roughly comparable in amplitude to that observed with the ENSO phenomenon (shown in Fig. 4.1).

© Springer International Publishing AG, part of Springer Nature 2018

I. Roy, *Climate Variability and Sunspot Activity*, Springer Atmospheric Sciences, https://doi.org/10.1007/978-3-319-77107-6_4

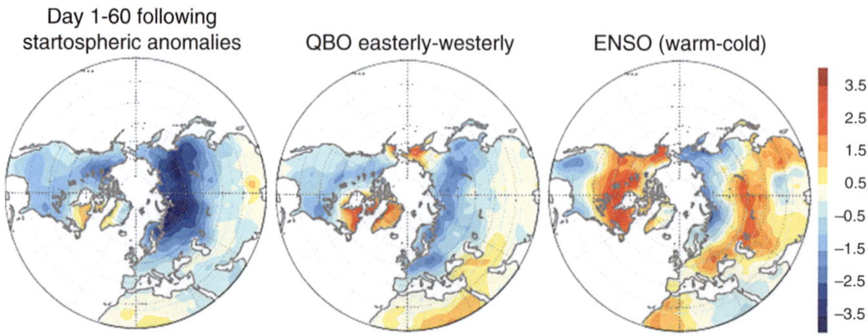

Fig. 4.1 The difference in mean surface temperature anomalies between the 60-day interval following the onset of strong and weak vortex events at 10 hPa level (left), during Januarys between QBO westerly and easterly (middle), and winters (Jan–Mar) for opposite phases of ENSO (right). Contour levels are at 0.5 C. (Source: Thompson et al. 2002, ©American Meteorological Society. Used with permission)

4.2 Polar Vortex and ENSO

Northern stratospheric polar vortex has been shown to be warmer and more perturbed during winters of El Niño years to that from La Niña. Camp et al. (2007) noted that during winter, warm-ENSO years are significantly warmer in the stratosphere at the midlatitudes and poles. Using GCM, Taguchi and Hartmann (2006) and Sassi et al. (2004) observed that the warming difference is statistically significant between La Niña and El Niño years. SSW is shown twice more likely to occur in the El Niño years than in La Niña winters. All those studies indicate a possible connection between the ENSO and polar stratosphere.

4.3 ENSO and Polar Troposphere

Some footprint of the ENSO in NAO is also observed by Toniazzo and Scaife (2006). They suggested that warm events of the ENSO are linked with the negative phase of the NAO. Of late, Ineson and Scaife (2009) using a general circulation model (GCM) detected a clear signature of the ENSO in European climate, and it is via the stratospheric pathway. The mechanism is restricted to years when SSW occurs. It leads to a transition to mild conditions in the Southern Europe and cold conditions in the Northern Europe during El Niño years in late winter. Those give an indication that there is a global scale teleconnection pattern. It involves the ENSO, where the role of the sun cannot be ignored.

The ENSO is likely to be associated with changes in the stratospheric Brewer–Dobson circulation with anomalous tropical lower stratospheric heating tending to strengthen the circulation and produce unusual adiabatic warming at high latitude, thus weakening the polar vortices (Haigh and Roscoe 2006). The imprint of ENSO in the stratosphere comprises of a warm and weak polar vortex in the Arctic stratosphere during El Niño. During the course of a winter, it propagates from the middle to the lower stratosphere. Thus, the ENSO via a downward propagation of strato- spheric anomalies might influence late winter weather (Randel et al. 2004). Such propagation of the ENSO signal is also reproduced by models (e.g. Manzini et al. 2006). They showed that the ENSO signal from the upper stratosphere in January travels to the lower stratosphere in February and March.

4.4   ENSO, Polar Annular Modes and JET

Recent studies have identified some signature of the ENSO in the polar modes of variability. Haigh and Roscoe (2006), using data from the latter half of the twentieth century, noticed a strong anti-correlation between the SAM and ENSO in the lower troposphere. Considering a period of 1979 to 2000, Carvalho et al. (2005) observed that during austral summer (DJF), cold ENSO events are associated with the domi- nant positive phase of AAO and vice versa for warm ENSO. The alternation of AAO phases was also shown to be allied with the variation in intensity of midlatitude polar jet (of ~ 60°S) and latitudinal migration of the upper level (200 hPa) STJ (of ~ 45°S). Positive (negative) phases of AAO are linked with the poleward (equatorward) shift and weakening (strengthening) of the subtropical feature accompanied by an intensi- fication (weakening) of the high latitude characteristic (shown in Fig. 4.2).

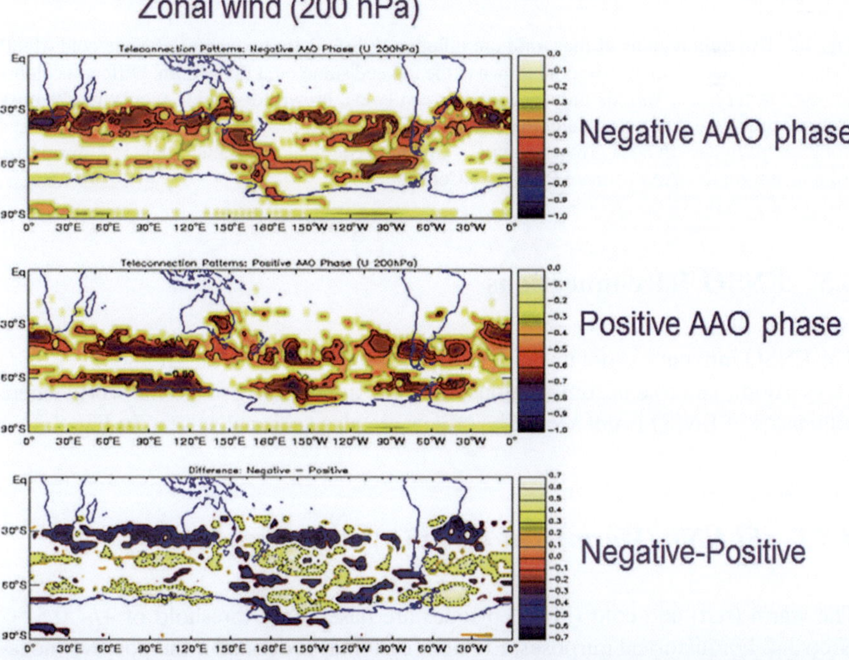

Fig. 4.2 Patterns of teleconnection observed for the zonal wind anomalies at 200 hPa level during negative (top) and positive (middle) AAO phase. Difference between these fields is presented at the bottom. (Carvalho et al. 2005, ©American Meteorological Society. Used with permission)

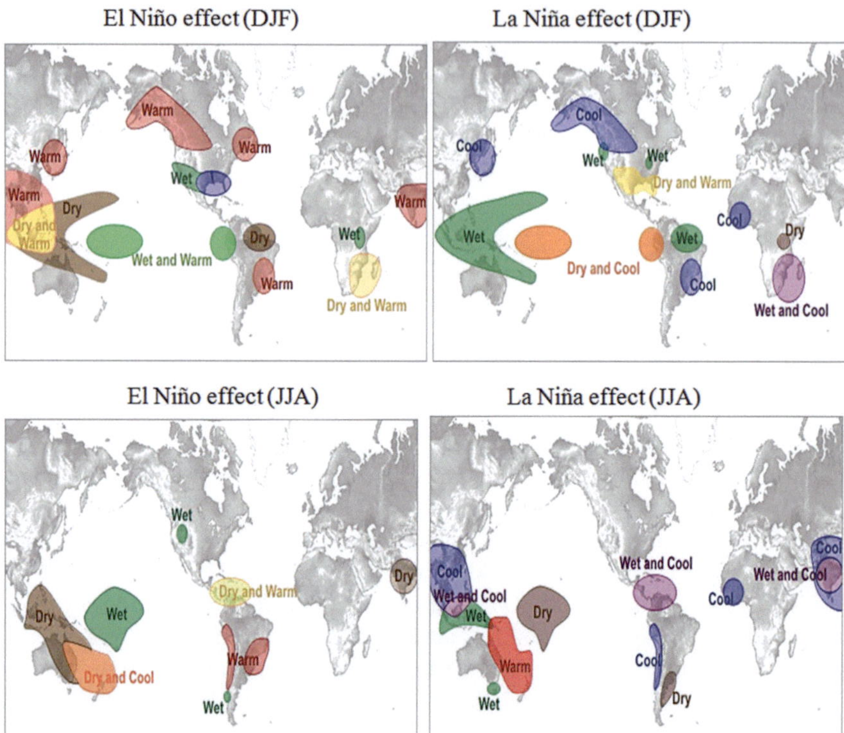

Fig. 4.3 Different regions of the world are influenced differently by warm (left) and cold (right) phases of ENSO. Northern winter is shown at the top and summer at the bottom. Different colours are used to represent various influences, e.g. green (wet), brown (dry), blue (cool), red (warm). (Source: https://climate.ncsu.edu/climate/patterns/ENSO.html, link accessed on 10/12/2017, used from http://www.srh.noaa.gov/jetstream/, Credit: National Oceanic and Atmospheric Administration (NOAA) Climate Prediction Center)

4.5   ENSO Teleconnections

The ENSO influence is not uniform and varies around the world. Though the ENSO plays a dominant role in some regions, but there are regions around the world where influences of ENSO is not seen. The effects are also seasonal as seen in Fig. 4.3.

4.5.1   *El Niño (Warm) and La Niña (Cold) Definition*

The warm (red) and cold (blue) episodes are based on a threshold of +/− 0.5 °C Niño 3.4. For historical purposes, El Niño (La Niña) is defined if the positive (negative) threshold is maintained for a minimum of five consecutive overlapping seasons as shown in Table 4.1. Here blue shows La Niña period while red El Niño.

Table 4.1 La Niña marked by blue and El Niño by red in recent years for different seasons

Year	DJF	JFM	FMA	MAM	AMJ	MJJ	JJA	JAS	ASO	SON	OND	NDJ
2010	1.5	1.3	0.9	0.4	-0.1	-0.6	-1.0	-1.4	-1.6	-1.7	-1.7	-1.6
2011	-1.4	-1.1	-0.8	-0.6	-0.5	-0.4	-0.5	-0.7	-0.9	-1.1	-1.1	-1.0
2012	-0.8	-0.6	-0.5	-0.4	-0.2	0.1	0.3	0.3	0.3	0.2	0.0	-0.2
2013	-0.4	-0.3	-0.2	-0.2	-0.3	-0.3	-0.4	-0.4	-0.3	-0.2	-0.2	-0.3
2014	-0.4	-0.4	-0.2	0.1	0.3	0.2	0.1	0.0	0.2	0.4	0.6	0.7
2015	0.6	0.6	0.6	0.8	1.0	1.2	1.5	1.8	2.1	2.4	2.5	2.6
2016	2.5	2.2	1.7	1.0	0.5	0.0	-0.3	-0.6	-0.7	-0.7	-0.7	-0.6
2017	-0.3	-0.1	0.1	0.3	0.4	0.4	0.1	-0.2	-0.4	-0.7		

Source: http://origin.cpc.ncep.noaa.gov/products/analysis_monitoring/ensostuff/ONI_v5.php, link on 11/12/2017, Credit: National Oceanic and Atmospheric Administration (NOAA) Climate Prediction Center

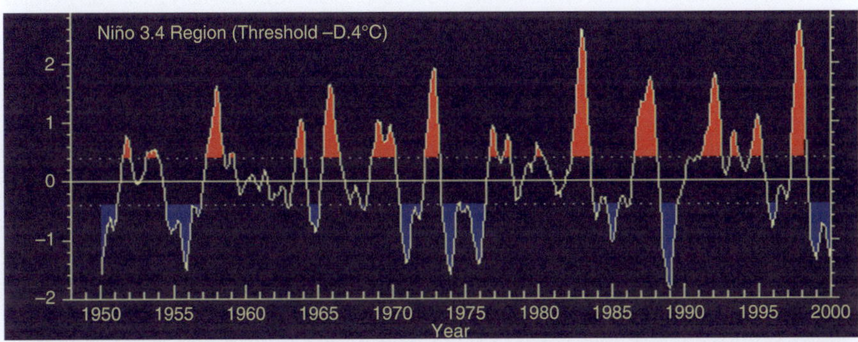

Fig. 4.4 Niño 3.4 time series demarcating threshold value (here 0.4 °C) for El Niño (red) and La Niña (blue). (Source: http://www.cgd.ucar.edu/cas/catalog/climind/Nino_3_3.4_indices.html, link accessed on 11/12/2017 shows weekly SST totals for the past (recent) 52 weeks), Animation http://www.esrl.noaa.gov/psd/map/clim/sst.anim.year.html, link accessed on 11/12/ 2017 shows weekly SST totals for the past (recent) 52 weeks

4.5.2   El Niño or La Niña?

Time series plot of Niño 3.4 SST in Fig. 4.4 indicates when it could be defined as El Niño or La Niña.

Animation [http://www.esrl.noaa.gov/psd/map/clim/sst.anim.year.html, link on 11/12/ 2017] shows weekly SST totals for the past (recent) 52 weeks.

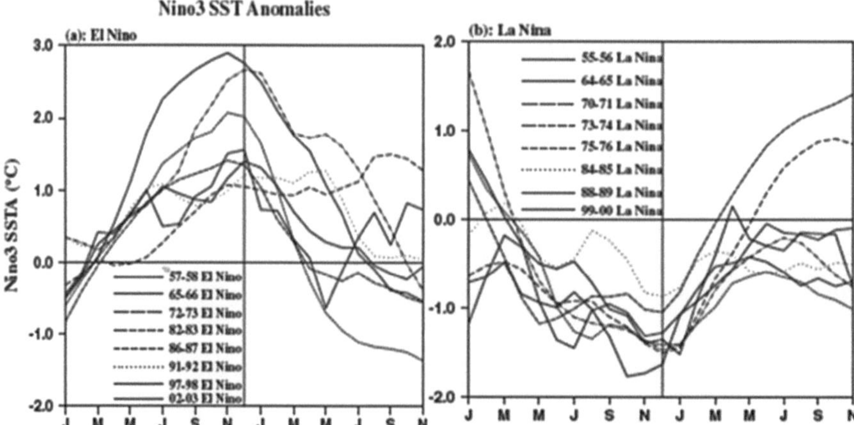

Fig. 4.5 SST anomalies from Jan of ENSO development year to Nov following year for (**a**) 8 El Niño warm event (**b**) 8 La Niña events during 1950–2003. Data NCEP monthly means SST (Wang and Fiedler 2006)

4.5.3   ENSO Seasonal Locking

One interesting characteristic of the ENSO is its phase locking with the season. The phenomenon usually peaks around the Christmas holidays, and hence the name El Niño was chosen which means the 'Christ Child'. The observations of Wang and Fiedler (2006) suggest that ENSO events tend to onset, grow and decay at certain seasons of the year (Fig. 4.5).

4.5.4   Potential Problems with SST Data

Vecchi et al. (2008) discuss how inconsistencies of the various SST reconstructions in the equatorial eastern Pacific indicate different results between the data from HadISST and NOAA. The two datasets differ in their analysis procedures, in data sources and the corrections algorithm incorporated before the 1940s. Regarding sources, the NOAA data consider only in situ measurements, while the Hadley Centre include satellite-derived SST data since the beginning of the 1980s. The latter also covers additional in situ observations collected from the UK Meteorological Office archive which are absent in the former (Vecchi and Soden 2007). They noted that linear trends in tropical Pacific SST over the period 1880–2005 exhibit a 'La Niña-like' pattern if computed using the HadISST reconstructions (Rayner et al. 2003). Whereas using the NOAA extended SST reconstruction (Smith and Reynolds 2004), it suggests an 'El Niño-like' signature (Fig. 4.6).

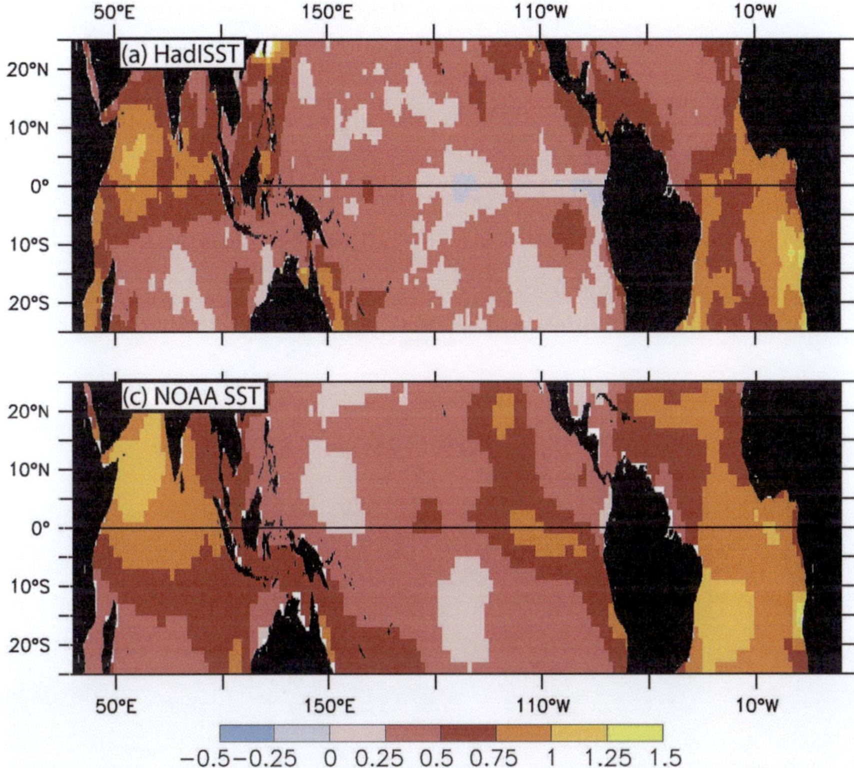

Fig. 4.6 The linear trend in historical SST anomaly (in °C between 1880 and 2005), from two different reconstructions: (**a**) Hadley Centre and (**b**) NOAA. Monthly mean climatology removed before computing trend. (Vecchi and Soden 2007, ©American Meteorological Society. Used with permission)

4.5.5   *Indian Summer Monsoon and Walker Circulation*

It is noticed that there are usually scanty of rainfall during ISM on the El Niño years. On the other hand, heavy rains are seen during La Niña years. The direction of change in Walker circulation could be one responsible factor. During El Niño years, Walker circulation is seen to be suppressing ISM rainfall as indicated with the dry descending branch close to Indian subcontinent (Fig. 4.7). The warm ocean water, around eastern tropical Pacific during El Niño, causes upward convection zone and rising branch of the Walker circulation in that place. However, during La Niña or non-El-Niño years, warm ocean waters accumulate in the western side of the tropical Pacific. Thus, the rising branch of Walker circulation accompanied by associated convection takes place around the west Pacific. Indian subcontinent being closer to that convection zone receives more moisture, which is also stronger during La Niña phase, and causes heavy rains.

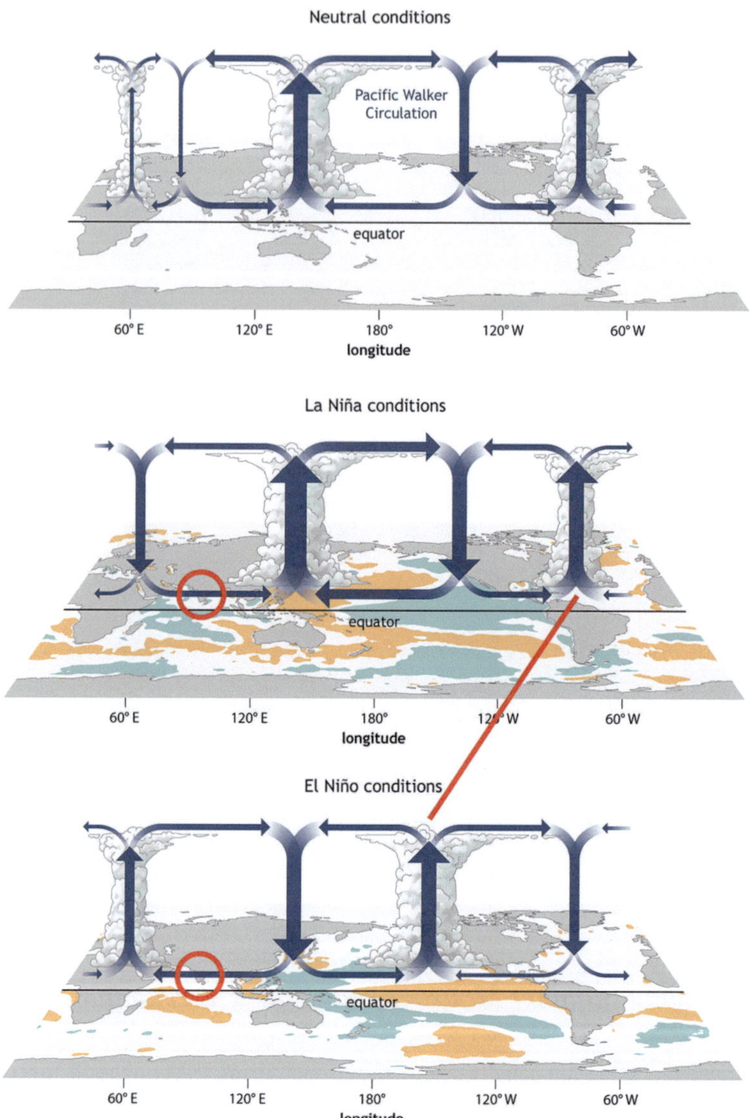

Fig. 4.7 Change in Walker circulation during El Niño to that from La Niña can impact Indian summer monsoon rainfall. (Source: https://www.climate.gov/sites/default/files/Walker_Neutral_large.jpg; https://www.climate.gov/sites/default/files/Walker_LaNina_2colorSSTA_large.jpg; https://www.climate.gov/sites/default/files/Walker_ElNino_2colorSSTA_large.jpg   Credit: National Oceanic and Atmospheric Administration (NOAA) Climate Prediction Center)

4.5.6   Different Types of ENSO

Recent studies suggested that the ENSO can be classified in various categories following their formation mechanism and specific characteristics. One pattern showed stronger SST features around central Pacific, often known as central Pacific (CP) type or Modoki, whereas other forms showed more robust feature around eastern Pacific, often known as east Pacific (EP) type or Canonical ENSO. Modoki is more atmosphere related and got a connection with midlatitudes, whereas canonical form is mainly associated with thermocline shifting, and ocean plays an important role.

For the purpose of definition, different regions of tropical Pacific are considered as regions A, B and C and canonic region and are shown in Fig. 4.8.

First, anomalous SSTs (SSTA) are calculated for four different regions: i) canonic (5°N-5°S, 90°W-140°W), ii) A (10°S-10°N,165°E-140°W), iii) B (15°S- 5°N, 110°W-70°W), and iv) C (10°S-20°N,125°E-145°E). Then ENSO Modoki index (EMI) is calculated using this relationship.

$$\boxed{\text{ENSO Modoki Index}\left(\textbf{EMI}\right) = \text{ISSTAI } \textbf{A} - 0.5^{*}\text{ISSTAI } \textbf{B} - 0.5^{*}\text{ISSTAI } \textbf{C}}$$

Three different types of ENSO are defined as follows based on the works of Kao and Yu (2009), Ashok et al. (2007) and Kug et al. (2009).

ENSO Modoki (**ENM or LNM**): EMI is greater than $0.7\sigma_M$; σ_M is Std. dev. of EMI. For warming, it is EN and cooling LN.

ENSO Canonical (**ENC or LNC**): If mean SSTA in canonical region is greater than $0.7\sigma_C$; σ_C is standard deviation of a season in that region.

ENSO Canonical and Modoki (**ENCM or LNCM**): ENSO satisfies both canonical and Modoki features.

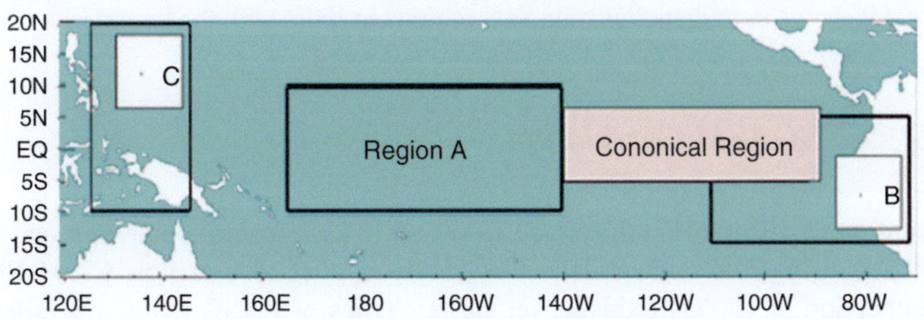

Fig. 4.8 Various regions of tropical Pacific considered for defining Canonic and Modoki ENSO (Roy et al. 2017)

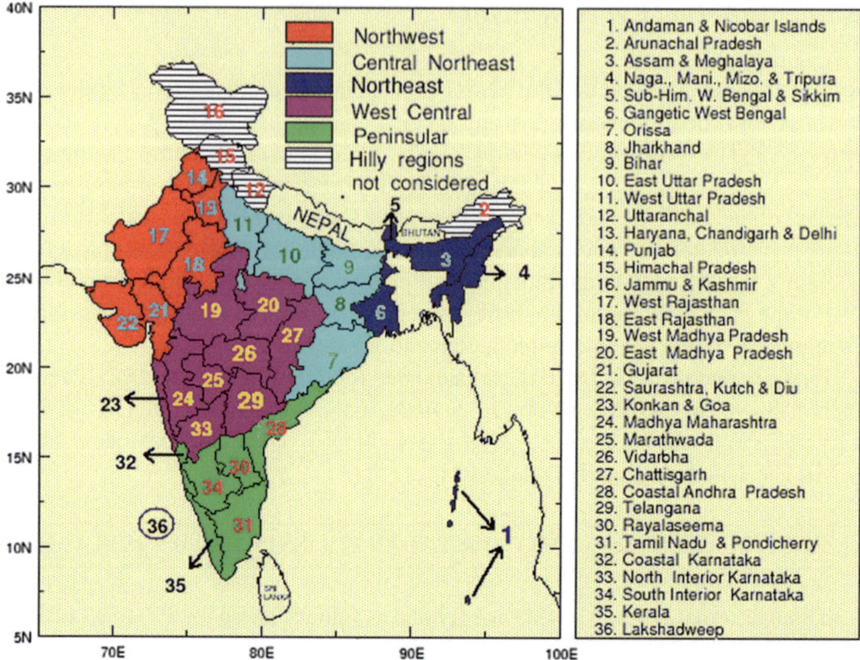

Fig. 4.9 Homogeneous monsoon regions in India, six different regions are marked by different colours. Numbers show various states in those regions, and their full forms are shown on the right-hand side. (Source: http://nihroorkee.gov.in/rbis/india_information/monsoon1.jpg, uploaded on 31/12/2017)

4.5.7   Homogeneous Monsoon Region

ISM rainfall is not uniform throughout the countries, and there are regional differences. Figure 4.9 shows homogeneous monsoon regions in India with the boundaries.

4.5.8   ENSO ISM Correlation

A recent paper (Roy 2017) discussed ISM and ENSO general behaviour and their correlation using different models. Few recent studies also explored ISM-ENSO teleconnection in models focusing on various types of ENSO (Roy and Tedeschi, 2016; Roy et al. 2017). Though the negative correlation between ISM and ENSO is commonly known, it is not present everywhere in Fig. 4.10. Many regions are showing no correlation at all. Canonic region and region A shows strong negative correlation around CNE India, parts of Hilly area and east coast of peninsular region (Fig. 4.10). Moreover, there is a positive correlation in central India for EMI.

But there is asymmetry in El Niño and La Niña mechanism, which indicates that correlation study can have certain limitations. There is also asymmetry in the spatial pattern of ENSO–ISM teleconnection in LN and EN phase. To overcome those drawbacks, compositing study can be useful.

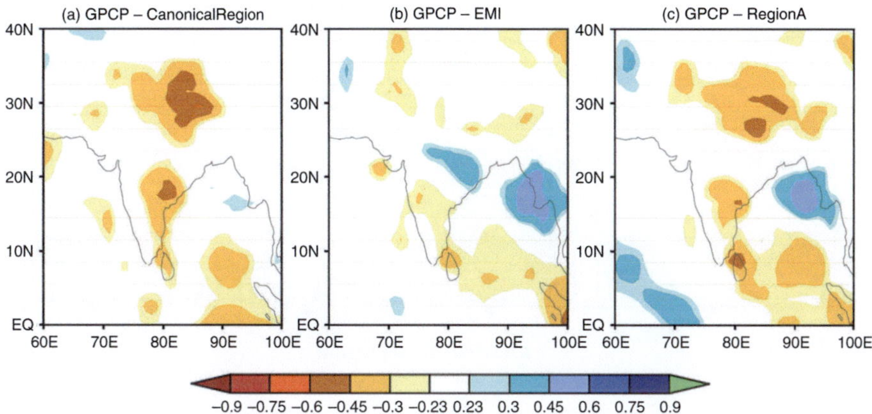

Fig. 4.10 ENSO ISM correlation is shown using canonical region (left), EMI (middle) and region A (right) using observation from GPCP rainfall data (Roy et al. 2017)

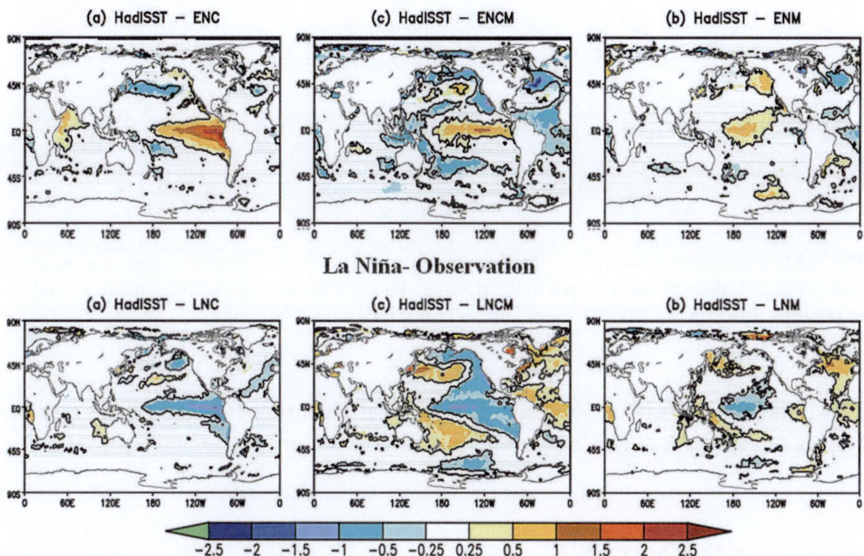

Fig. 4.11 SST composites for El Niño (top) vs. La Niña (bottom) using HadISST data. Modoki phase is shown in the right (**b**), canonic phase left (**a**), with Canonic–Modoki in between (**c**) (Roy et al. 2017)

4.5.9   SST Composites: EN vs. LN

Figure 4.11 shows that the spatial pattern of El Niño and La Niña differs in respective compositing for SST. LNCM is more extensively covered along the west coast of America, showing cold temperature all along, while such feature is not seen for ENCM. Opposite signature in tropical Pacific and Atlantic is noticed for ENM/

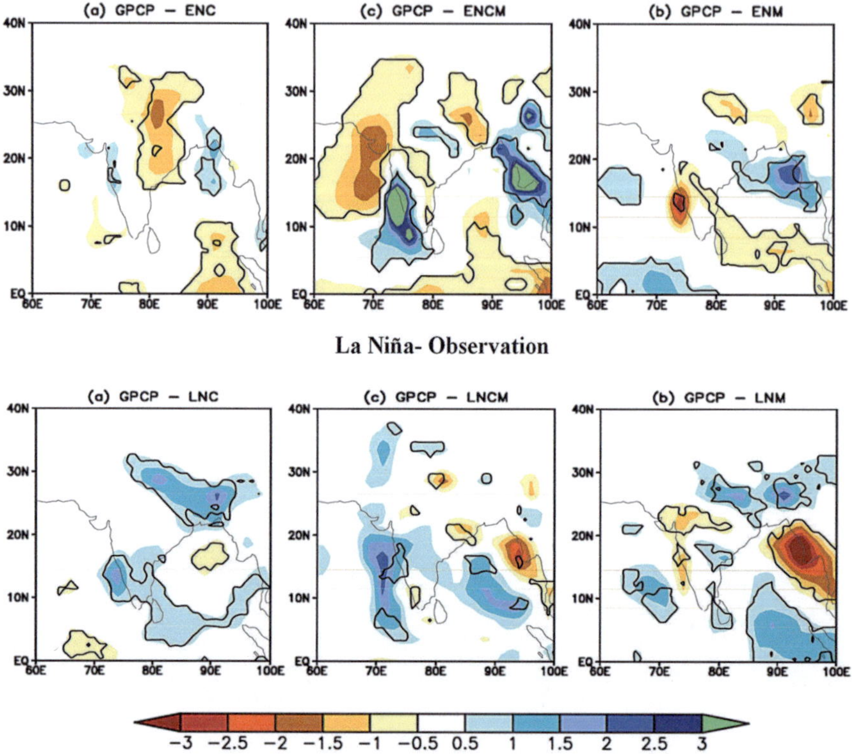

Fig. 4.12 Precipitation composites for El Niño (top) vs. La Niña (bottom) using observational GPCP data. Modoki phase is shown in the right (**b**), canonic phase left (**a**), with Canonic–Modoki in between (**c**) (Roy et al. 2017)

LNM and ENCM/LNCM (extensive in tropical Atlantic). Reverse signal outside tropics of Pacific as horseshoe pattern is mainly seen for LNCM and ENCM/ENC.

4.5.10   ISM ENSO Teleconnection Compositing: EN vs. LN

Precipitation teleconnection in observation is not opposite for El Niño to that from its La Niña counterparts (Fig. 4.12) and indicates why composting method has advantages than that from correlation study.

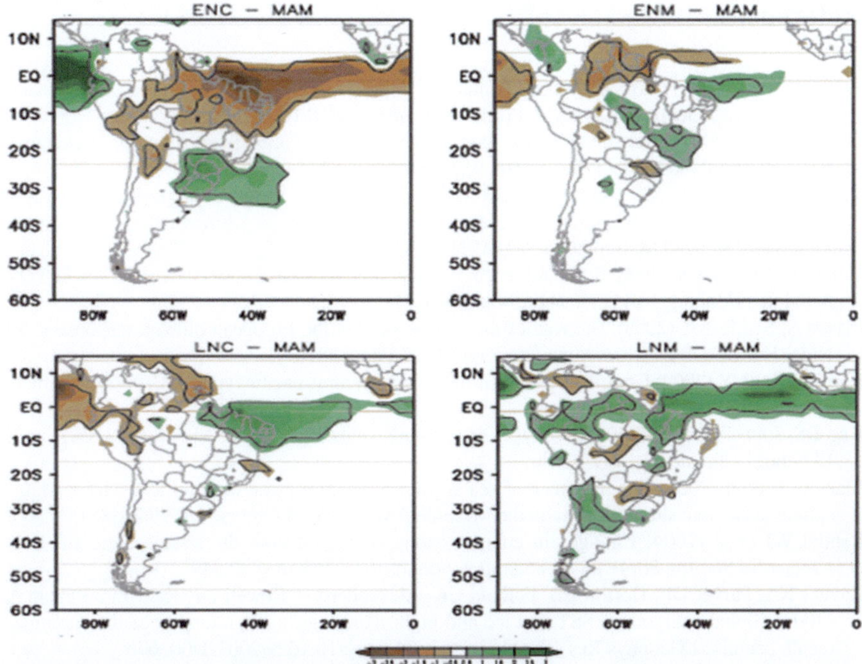

Fig. 4.13 Rainfall in different regions of South America is affected differently during Modoki (right) and Canonical phase (left) of ENSO during March, April and May. The top panel is for El Niño phase and bottom for La Niña (Tedeschi et al. 2013)

4.5.11   Rainfall in South America ENSO (Different Types) Teleconnection

Figure 4.13 shows composite in rainfall around South America during March, April and May for canonical and Modoki situation. It clearly suggests that the composites in La Niña are not the exact opposite of its El Niño counterparts. It also indicates there are plenty of differences in rainfall pattern during the Modoki phase to that from Canonical phase.

4.5.12   Summary: ENSO and Teleconnections

The ENSO being one of the major tropospheric variabilities, it is important to know about ENSO-related teleconnections. The topics covered here are i) ENSO, polar vortex and QBO, ii) ENSO and jet, iii) influence of ENSO around the world, iv) ENSO and ISM, v) various types of ENSO including Modoki and canonical type and vi) specific influences of east Pacific type (Canonic) and central Pacific type (Modoki). In the following chapter, solar influences around various places will be covered.

References

Ashok K, Behera SK, Rao SA, Weng H, Yamagata T (2007) El Niño Modoki and its possible tele-connections. J Geophys Res 112:C11007. https://doi.org/10.1029/2006JC003798

Camp CD et al (2007) Stratospheric polar warming by ENSO in winter: a statistical study. Geophys Res Lett 34:L04809. https://doi.org/10.1029/2006GL028521

Carvalho LMV, Jones C, Ambrizzi T (2005) Opposite phases of the Antarctic Oscillation and relationships with intraseasonal to interannual activity in the tropics during the austral summer. J Climate 18:702–718. https://doi.org/10.1175/JCLI-3284.1

Haigh JD, Roscoe HK (2006) Solar influences on polar modes of variability. Meteorol Z 15(3):371–378. https://doi.org/10.1127/0941-2948/2006/0123

Ineson S, Scaife AA (2009) The role of the stratosphere in the European climate response to El Niño. Nat Geosci 2:32–36. https://doi.org/10.1038/ngeo381

Kao H-Y, Yu J-Y (2009) Contrasting eastern-pacific and central-pacific types of El Niño. J Climate 22:615–632

Kug J-S, Jin F-F, An S-I (2009) Two types of El Niño events: cold tongue El Niño and warm pool El Niño. J Climate 22:1499–1515

Manzini E et al (2006) The influence of sea surface temperatures on the northern winter strato-sphere: ensemble simulations with the MAECHAM5 model. J Climate 19:3863–3881

Randel WJ et al (2004) Changes in column ozone correlated with the stratospheric EP flux. J Meteorol Soc Jpn 80(4B):849–862. https://doi.org/10.2151/jmsj.80.849

Rayner NA, Parker DE, Horton EB, Folland CK, Alexander LV, Rowell DP, Kent EC, Kaplan A (2003) Global analyses of SST, sea ice and night marine air temperature since the late nine-teenth century. J Geophys Res 108:4407. https://doi.org/10.1029/2002JD002670

Roy I (2017) Indian summer monsoon and El Niño southern oscillation in CMIP5 models: a few areas of agreement and disagreement. Atmosphere 8(12):154

Roy I, Tedeschi RG (2016) Influence of ENSO on regional ISM precipitation – local atmospheric Influences or remote influence from Pacific. Atmosphere 7:25. 10.3390/atmos7020025

Roy Y, Tedeschi RG, Collins M (2017) ENSO teleconnections to the Indian summer monsoon in observations and models. Int J Climatol 37(4):1794–1813. https://doi.org/10.1002/joc.4811

Sassi F et al (2004) Effect of El Nino-Southern Oscillation on the dynamical, thermal, and chemi-cal structure of the middle atmosphere. J Geophys Res 109:D17108

Smith TM, Reynolds RW (2004) Improved extended reconstruction of SST (1854–1997). J Clim 17:2466–2477. https://doi.org/10.1175/1520-0442. 017<2466:IEROS>2.0.CO;2

Taguchi M, Hartmann DL (2006) Increased occurrence of stratospheric sudden warmings during El Nino as simulated by WACCM. J Climate 19:324–332

Tedeschi RG et al (2013) Influences of two types of ENSO on South American precipitation. Int J Climatol 33:1382–1400. https://doi.org/10.1002/joc.3519

Thompson DWJ, Baldwin MP, Wallace JM (2002) Stratospheric connection to Northern hemi-sphere wintertime weather: implications for prediction. J Climate 15(12):1421–1428

Toniazzo T, Scaife AA (2006) The influence of ENSO on winter North Atlantic climate. Geophys Res Lett 33:L24704. https://doi.org/10.1029/2006GL027881

Vecchi GA, Soden BJ (2007) Global warming and the weakening of the tropical circulation. J Climate 20:4316–4340

Vecchi GA, Clement A, Soden BJ (2008) Examining the tropical pacific's response to global warming. Eos Trans AGU 89(9):81–83. https://doi.org/10.1029/2008EO090002

Wang C, Fiedler P (2006) ENSO variability and the eastern tropical pacific: a review. Prog Oceanogr 69(2006):239–266

Chapter 5
Solar Influence Around Various Places: Robust Solar Signal on Climate

Abstract This chapter focused on the detected robust solar signal on climate. It presented some observational results that identified solar signature on sea level pressure (SLP), sea surface temperature (SST) and annual mean air temperature. A technique of Multiple Linear Regression (MLR) methods was discussed, and a detected significant signal on SLP around Aleutian Low (AL) was analysed. A solar signal was observed around AL and Pacific High also using other techniques. In terms of SST, the region of tropical Pacific was addressed. One study noted an in-phase relationship between the Sun and tropical Pacific SST, and another study even observed a phase locking between those. A widely debated study that used the method of solar maximum compositing on tropical Pacific SST was presented discussing the methodology.

Keywords Sea Level Pressure · Sea Surface Temperature · Aleutian Low · Pacific High · Multiple Linear Regression · Compositing Technique · Solar Peak Year Compositing · Centre of Action

5.1 Signal on Sea Level Pressure (DJF) Using Multiple Linear Regression

Figure 5.1, using the MLR technique, suggested a strong solar signature on SLP during DJF around places of Aleutian Low (AL) with a secondary opposite signature around areas in the tropical Pacific. Significant signal around tropical Pacific can be responsible for inciting trade wind. Coincidentally, the location of PDO and ENSO (manifest as SST) also matches those places with their cold phases having same signed signature to that of high solar years. Signal around Aleutian Low is robust and insensitive to various methodologies and different time periods.

© Springer International Publishing AG, part of Springer Nature 2018

I. Roy, *Climate Variability and Sunspot Activity*, Springer Atmospheric Sciences, https://doi.org/10.1007/978-3-319-77107-6_5

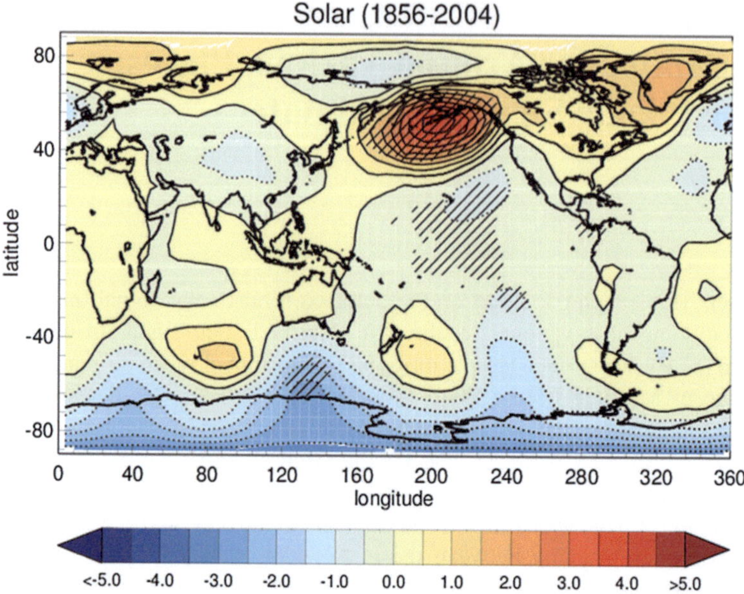

Fig. 5.1 Amplitudes of the components of SLP variability due to the Sun (monthly sunspot number) during DJF (in hPa) using Multiple Linear Regression (MLR). Other independent parameters considered are OD, trend and ENSO. Dashed lines suggest negative values, while hatching shows areas which are statistically significant at the 5% level (Roy 2014)

5.1.1 Method of Multiple Regression Analysis

Multiple linear regression may be represented as

$$y(x,t) = \Sigma \beta_i(x) f_i(t) + \text{noise}$$

where:

- $y(x,t)$ are data; here it is SLP or SST.
- $\beta_i(x)$ is the weight of contributing factor i at point x.
- $f_i(t)$ is time-dependent climate factor i. Usually, those are a linear trend, optical depth (OD) to represent volcano, solar cycle variability (here SSN) and ENSO as shown in Fig. 5.2.
- y and f are known; β are to be estimated.
- Noise is represented by an AR(1) model.

Here amplitudes of variability due to various climate factors are estimated using autoregressive noise model order one (AR(1)). An amplitude of variability due to SSN is calculated, to detect solar signal on SLP or SST. In this methodology, noise coefficients are calculated simultaneously with the components of variability. It is done in a manner so that the residual is consistent with an order one red noise model. This technique is also applied in recent studies to detect solar signal on surface climate (Roy and Collins 2015; Roy et al. 2016; Roy 2018).

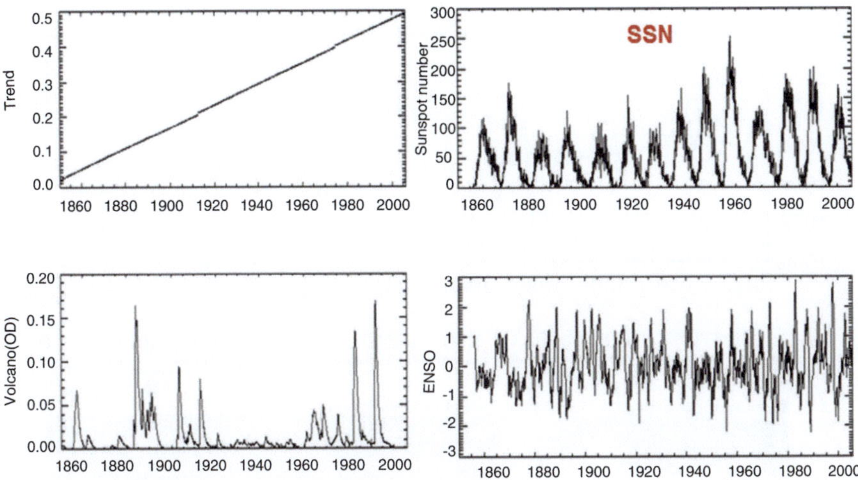

Fig. 5.2 Time series of trend, volcano, sunspot number (SSN) and ENSO

5.2   Solar Signal Around Aleutian Low (AL) and Pacific High (PH)

Christoforou and Hameed (1997) showed that during sunspot maxima, the centre of action (COA) of the Aleutian Low (AL) pressure system moves north westward (the movement to the west by as much as 700 km). On the other hand, the COA of the North Pacific High (PH) pressure system moves north by as much as 300 km. During a sunspot minima, the COA for these pressure systems moves in the reverse direction (Fig. 5.3b). Shifts in the COA change storm trajectories and cause signifi-cant anomalies in regional climate. Apart from varying position, AL even exhibits large changes in intensity; there is the average 1.6 mb difference between years of extreme solar activity (Fig. 5.3a).

5.3   Solar Influence: Tropical Pacific SST

The existence of the 11-year solar cycle signals in meteorological fields of the lower atmosphere/upper ocean has been detected in various observational analyses (Haigh 2003; Van Loon et al. 2007). However, the response found is latitudinal non-uniform and larger than would be expected from radiative forcing considerations alone. White et al. (1997), using EOF analysis of SST data, during the second half of the twentieth century, detected a leading mode of variability. Its spatial pattern suggests a warming around the eastern Pacific in the tropics which is accompanied by a cool-ing centred around the North Pacific (160°W and 30°N) (Fig. 5.4). The 11-year

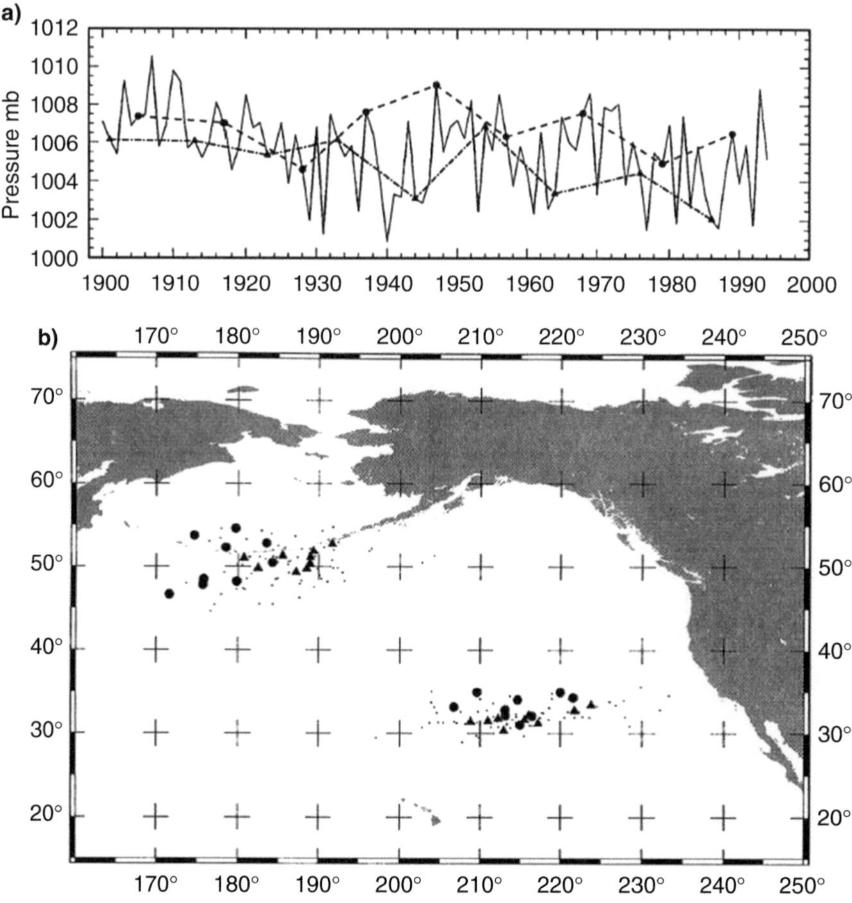

Fig. 5.3 *Pressure* around AL deviates between years of low and high solar activity (**a**), along with the *locations* of AL and PH (**b**); triangles indicate years of low solar activity and circles years of high solar activity. (Source: Christoforou and Hameed 1997)

period is shown as lagging the solar cycle by ~1 year (top panel). The upper panel in that figure indicates the time-varying amplitude, whereas the lower panel indicates the pattern of response. Such a spatial pattern is clearly a reminiscence of the warm event of ENSO.

5.4 ENSO and Sun Phase Locking

The recent study by White and Liu (2008), using the method of singular value decomposition (SVD) and compositing of nine solar cycles (period 1900–2000), even detected the phase locking of harmonics of the ENSO time series and the solar cycle.

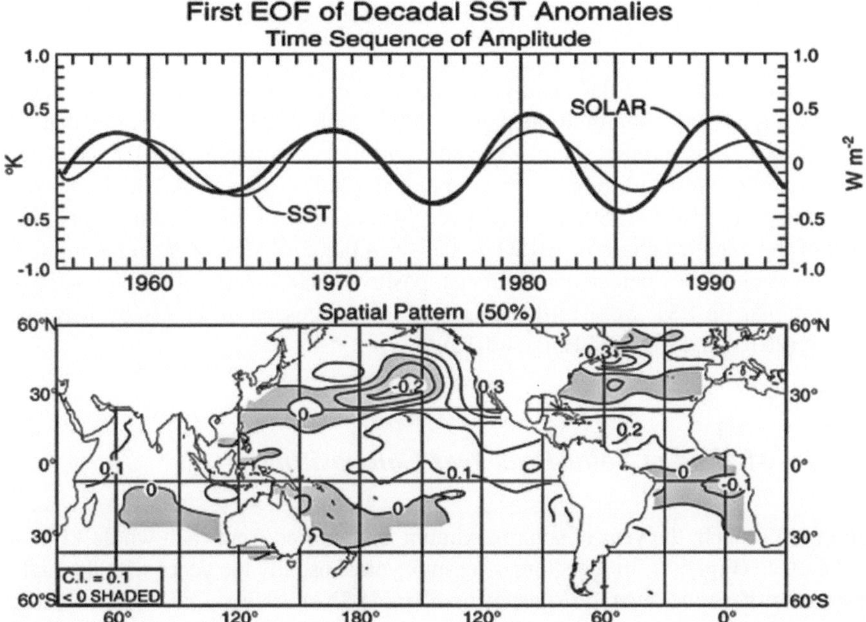

Fig. 5.4 Time series (top) of the magnitude and a spatial pattern (bottom) of response in SST associated with solar variability using EOF analysis for a period: 1955–1994 (White et al. 1997)

Fig. 5.5 Phase locking of quasi-decadal solar oscillation (QDO) (grey line) and sum of ENSO signals (thick line) (White and Liu 2008)

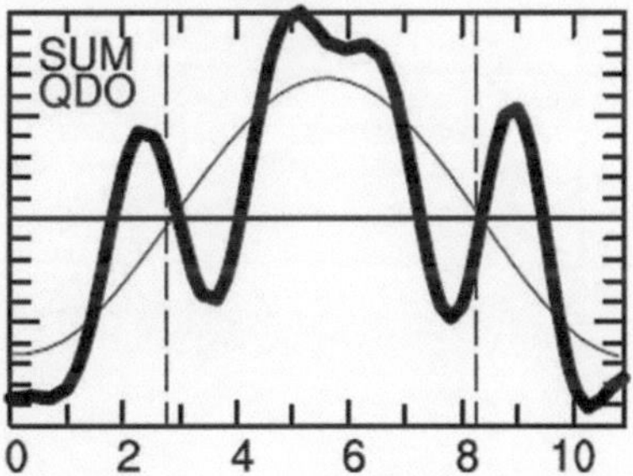

It resulted in a warm phase-like signature for about 3 years around the peak of the decadal solar oscillation (DSO), whereas cold phases lie approximately 2 years either side of the peak. The stronger warm phases peak 4–3 years after and before it (shown in Fig. 5.5).

5.5   Solar Signal in Tropical Pacific SST Using Compositing

Using the method of solar peak year compositing (of nearly 150 years of data), Meehl and co-authors in multiple papers (2007, 2008, 2009) have indicated that for an enhance solar forcing, there is a cold event like situations in the Pacific. They considered the season DJF only. Van Loon et al. (2007) using NOAA SST datasets, over the period 1871–1989, detected a very strong tropical solar signal, resembling that of the negative phase of ENSO as shown in Fig. 5.6. Moreover, in the midlatitude of Pacific, they detected a significant positive solar fingerprint. Their results of solar signal on SST were further analysed and discussed in details by Roy and Haigh (2010) and Roy and Haigh (2012).

5.5.1   Method of Solar Peak Year Compositing

In this approach, the year of solar maximum is identified from each whole 11-year solar cycle (Fig. 5.7). In detecting solar max years, usually the years with the highest value of annual average sunspot numbers (SSNs) are chosen. Then the mean

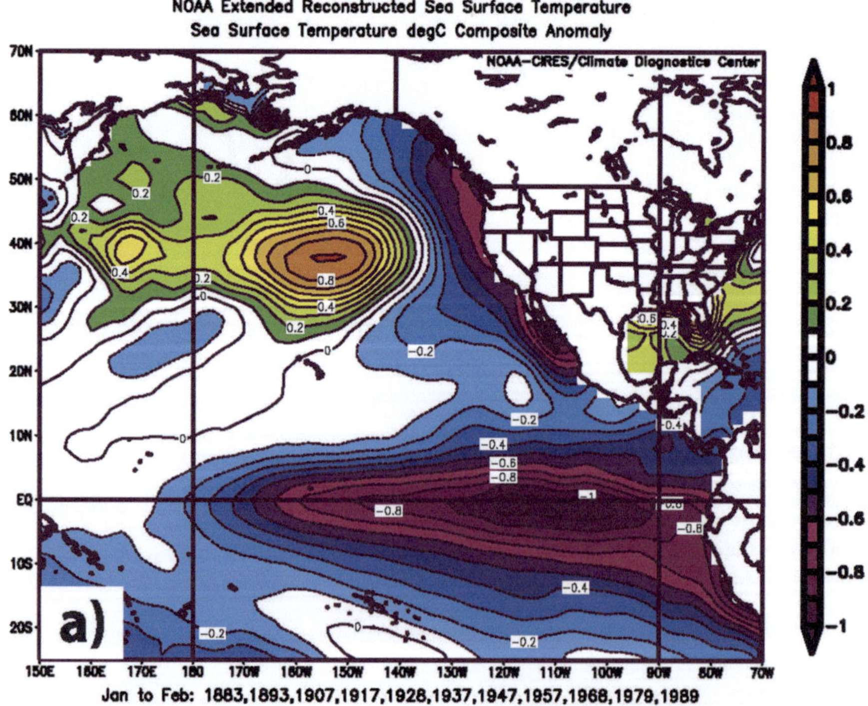

Fig. 5.6 Solar signal for SST using solar max compositing in °K using NOAA SST data (Van Loon et al. 2007)

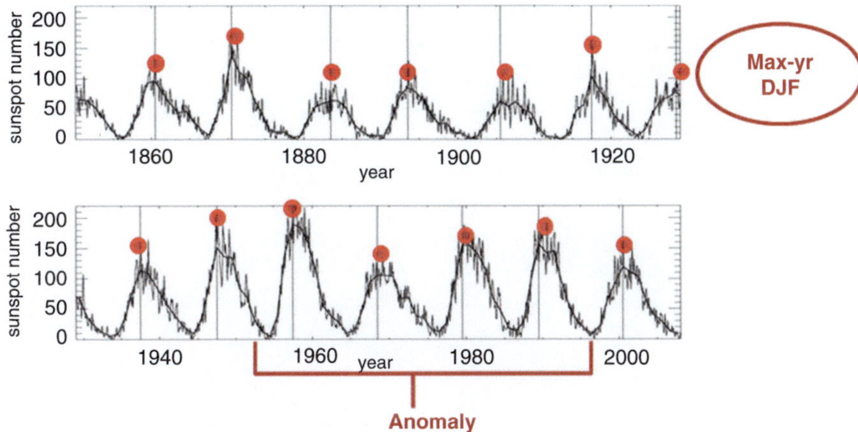

Fig. 5.7 Schematic representation of method of solar peak year compositing (during DJF here) as is considered by van Loon and Meehl in various studies

value, over all the peak solar years, is calculated for each meteorological parameter of interest (here, SLP or SST). The anomaly of that mean value about some reference level is calculated and considered as the response of that parameter to enhanced solar activity. The robustness of the signal is then examined through significant testing.

5.6   Observation: Annual Mean Temperature

Using rocketsonde data, Lidar data, model assimilation dataset, satellite data from stratospheric sounding unit (SSU) and microwave sounding unit (MSU), several analyses were carried out to find the solar influence on the middle atmosphere (Keckhut et al. 2004; Hood 2004). The signal in temperature as observed by Keckhut et al. (2004) is shown in Fig. 5.8(top). Using a fixed dynamical heating (FDH) model, Gray et al. (2009) were able to demonstrate that the lower stratospheric temperature signal is a response consistent with the change in ozone at those levels. It is seen that apart from a maximum in the upper stratosphere, there also exist a secondary peak in the lower stratosphere in the signals in temperature (Fig. 5.8). There is a consensus regarding the distinct solar signature of a 11-year cycle in stratospheric temperature; however, the discrepancies in the detailed analysis of the amplitudes and distribution of the detected signals, as seen in these figures, still need to be resolved.

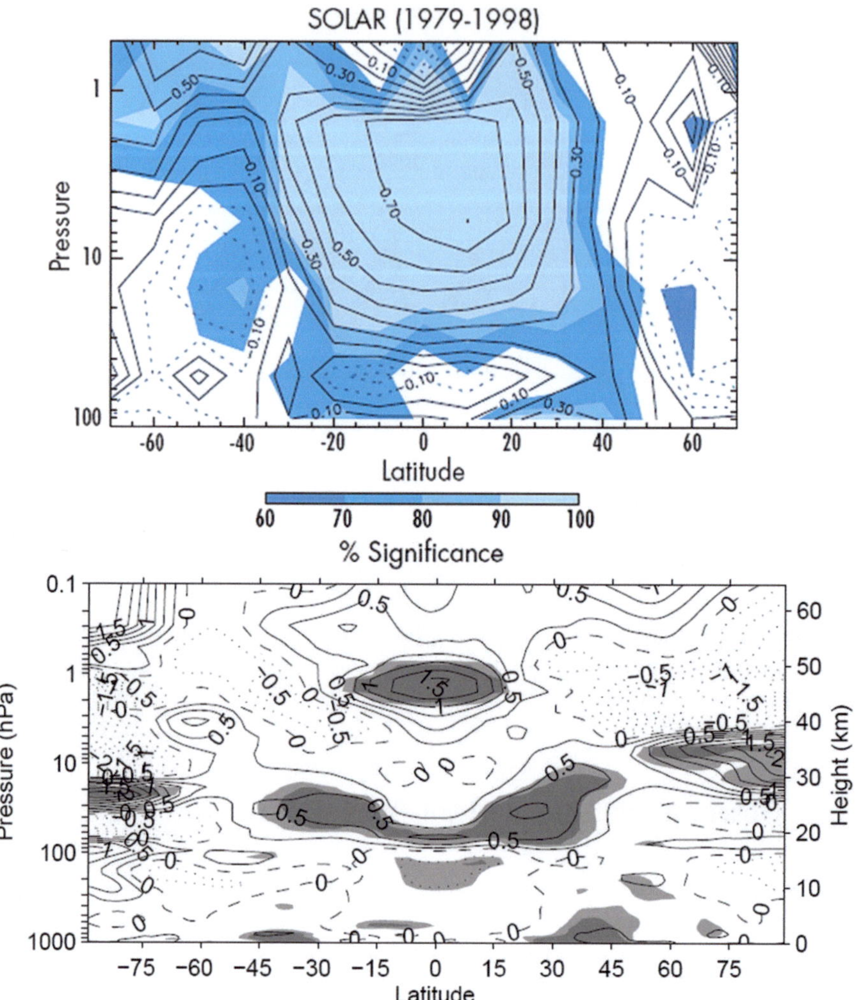

Fig. 5.8 Annual mean temperature difference (in °K) during solar max to solar min from observational data: (**a**) SSU/MSU satellite data analysis by Keckhut et al. (2004) for period 1979–1998 with various levels of statistical significance; (**b**) ERA reanalysis data for period 1978–2008 with dark/light shading indicates 99%/95% significance level (Frame and Gray 2010)

References

Christoforou P, Hameed S (1997) Solar cycle and the Pacific 'centers of action'. Geophys Res Lett 24(3):293–296. https://doi.org/10.1029/97GL00017

Frame THA, Gray LJ (2010) The 11-year solar cycle in ERA-40 data: an update to 2008, J. Clim, Early online release. https://doi.org/10.1175/2009JCLI3150.1

Gray LJ, Rumbold ST, Shine KP (2009) Stratospheric temperature and radiative forcing response to 11-year solar cycle changes in irradiance and ozone. J Atmos Sci 66(8):2402–2417

Haigh JD (2003) The effects of solar variability on the Earth's climate. Philos T R Soc A 361(1802):95–111

Hood LL (2004) Effects of solar UV variability on the stratosphere, solar variability and its effects on climate. Geophys Monogr Amer Geophys Union 141:283–303

Keckhut P, McDermid S, Swart D et al (2004) Review of ozone and temperature lidar validations performed within the framework of the network for the detection of stratospheric change. J Environ Monit 6(9):721–733

Meehl GA, Arblaster JM (2009) A lagged warm event-like response to peaks in solar forcing in the Pacific region. J Clim 22(13):3647–3660

Meehl GA, Arblaster JM, Branstator G, van Loon H (2008) A coupled air-sea response mechanism to solar forcing in the Pacific region. J Climate 21(12):2883–2897

Roy I (2014) The role of the Sun in atmosphere-ocean coupling. Int J Climatol 34(3):655–677

Roy (2018) Solar cyclic variability can modulate winter Arctic climate. Scientific Reports 8 (1)

Roy I, Haigh JD (2010) Solar cycle signals in sea level pressure and sea surface temperature. Atmos Chem Phys 10(6):3147–3153

Roy I, Haigh JD (2012) Solar cycle signals in the Pacific and the issue of timings. J Atmos Sci 69(4):1446–1451

Roy I, Collins M (2015) On identifying the role of Sun and the El Niño southern oscillation on Indian summer monsoon rainfall. Atmos Sci Lett 16(2):162–169

Roy I, Asikainen T, Maliniemi V, Mursula K (2016) Comparing the influence of sunspot activity and geomagnetic activity on winter surface climate. J Atmos Sol Terr Phys 149:167–179

van Loon H, Meehl GA, Shea DJ (2007) Coupled air-sea response to solar forcing in the Pacific region during northern winter. J Geophys Res-Atmos 112:D02108. https://doi.org/10.1029/2006JD007378

White WB, Liu ZY (2008) Non-linear alignment of El Nino to the 11-yr solar cycle. Geophys Res Lett 35:L19607. https://doi.org/10.1029/2008GL034831

White WB, Lean J, Cayan DR, Dettinger MD (1997) Response of global upper ocean temperature to changing solar irradiance. J Geophys Res-Oceans 102(C2):3255–3266

Chapter 6
Total Solar Irradiance (TSI):
Measurements and Reconstructions

Abstract This chapter focuses on total solar irradiance (TSI). It gives a very brief idea about their measurements and reconstructions. It also discussed different TSI, e.g. Solanki and Krivova, Foster and Hoyt and Schatten.

Keywords Total Solar Irradiance (TSI) · Hoyt and Schatten TSI · Foster's TSI · PMOD Composites

Direct measurements of TSI from outside the earth were initiated in 1978 with the advent of satellite instruments. As the current satellite era covers information only on the short-term components of solar variability, to assess the potential influence on a long-term basis, it is important to know TSI further back. In reconstructing past changes in TSI, proxy indicators of solar variability are considered to generate an estimate of its temporal variation over the past centuries. As for this, longer periods of observation are available.

However, the satellite data for recent decades suggests that significant uncertainties are present in all the existing satellite measurements (as shown in Fig. 6.1). It is relating to calibration of the instruments and degrading over time. It shows an attempt at compositing the measurements (shown in lower panel of Fig. 6.1) that produce the best estimate (Claus Fröhlich: http://www.pmodwrc.ch/).

Due to the problem of unexplained drift and uncalibrated degradation in the time series of the data obtained from satellite, there is an additional uncertainty, in the existence of any underlying trend in TSI over the last two cycles. The compositing as shown in Fig. 6.1 suggests that between the cycle minima occurring in 1986 and 1996, there is essentially no difference in TSI values. However, the results of Willson and Mordinov (2003) find an increase in irradiance (~0.045%) between these two consecutive solar cycle minima. Such discrepancies are important because all the available TSI reconstructions rely heavily on satellite data.

There are several different approaches adopted for reconstructing the TSI, and all those employ a substantial degree of empiricism. The first method uses SSN for TSI reconstruction (Solanki and Krivova 2003; Hoyt and Schatten 1993; Foster 2004; Lean et al. 1995). The second approach uses aa geomagnetic records

© Springer International Publishing AG, part of Springer Nature 2018

I. Roy, *Climate Variability and Sunspot Activity*, Springer Atmospheric Sciences, https://doi.org/10.1007/978-3-319-77107-6_6

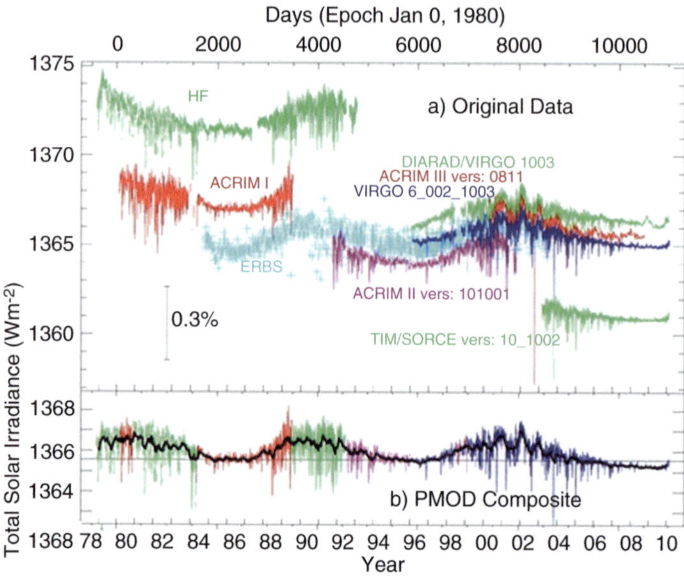

Fig. 6.1 Total solar irradiance daily average: top) measurement from various satellites; bottom) best estimate of TSI using compositing method. (Source: https://www.pmodwrc.ch/forschungent-wicklung/solarphysik/tsi-composite/, link accessed dt 2/4/18, credit Claus Fröhlich)

(Lockwood et al. 1999), while the third one involves using climate records (Reid 1997; Beer et al. 2000). The final approach for TSI reconstruction uses numerical models of the Sun to simulate the variations in solar parameters (Sofia and Li 2001). The review by Gray et al. (2005) discusses the various approaches.

In most SSN-based methods, the solar radiative output is calculated by a balance between decrease due to the presence of SSNs and increase due to the development of faculae, the bright patches on the Sun's surface. The darkening of sunspot depends on the area of the Sun covered by the SSN. Whereas the facular brightening relates to some indices, which might include solar cycle length, SSN (Lean et al. 1995), solar rotation rate, solar cycle decay rate and various empirical combinations of all these (Hoyt and Schatten 1993; Solanki and Fligge 1998). In determining long-term variability, different techniques were employed in various reconstructions. For example, Lean et al. (1995) used sunspot cycle amplitude to determine long-term variability, whereas Hoyt and Schatten (1993) used mainly sunspot length. Solanki and Fligge (1998) made an additional assumption considering the long-term contri-bution of the 'quiet sun'. It is followed by observations of Sun-like stars (Baliunas and Jastrow 1990) along with the assumption that Sun during the Maunder Minimum was in a non-cycling state.

Considering longer time series of this reconstruction based on data from Lean (2000), Wang et al. (2005) and Foster (2004) as shown in Fig. 6.2, it is obvious that the estimates diverge when it go back in time. Hence, over this longer period, the value of solar radiative forcing is highly uncertain. Moreover, this plot of Fig. 6.2 does not include any additional estimates which may suggest a much larger variation

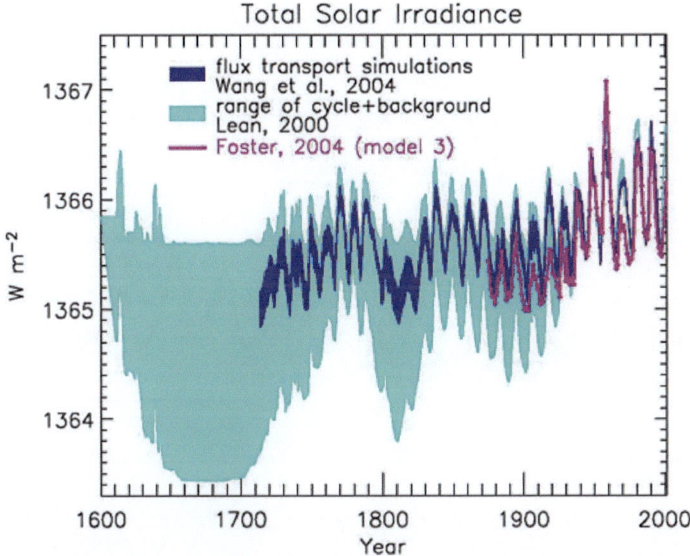

Fig. 6.2 Reconstructions of total solar irradiance since 1600 by various authors (Lean et al. 2005)

between the present level and that occurred during the Maunder Minimum in SSN (e.g. Hoyt and Schatten 1998) which prevailed during the seventeenth century (latter part). Such uncertainties are likely to affect studies relating to solar-climate research that considers very longer time record.

The time series of three other TSI, viz. Solanki and Krivova, Foster and Hoyt and Schatten, are shown in Fig. 6.3. The TSI of Solanki and Krivova is mainly based on facular brightening and relative SSN (Solanki et al. 2003). Whereas Foster's TSI is largely based on observation on sunspots (Foster, thesis, 2004). The TSI from Hoyt and Schatten primarily considers solar cycle length and thus mainly captures long-term solar variations rather than the 11-year solar cycle variability. It can be found from http://climexp.knmi.nl/data/isolarconstanths.dat, also in Hoyt and Schatten (1993).

Several earlier studies mentioned about uncertainty in using appropriate TSI in climate research (Mann et al. 2005; Brönnimann et al. 2005). Stott et al. (2001) discussed the differences for their model experiments using two solar forcing dataset, Lean et al. (1995) and Hoyt and Schatten (1993). The Hoyt and Schatten forcing is a closer fit of the model response to the noted warm maximum that happened in the 1940s (Smith et al. 2003; Stott et al. 2001). Apart from TSI the other measures of solar activity which include SSN, solar flux at 10.7 cm, aa index, etc. also vary widely. For example, F10.7 cm flux and SSN reached their peak values over the last century at the solar max of 1958, whereas the aa index showed its highest value during 1990. It is because the aa index shows the 11-year cycle imposed on a longer-term modulation, whereas the SSN return to virtually zero at each solar minimum (Lean and Rind 1998). Such differences between the solar forcing datasets are essential for detection/attribution of climate forcing parameters (Meehl 2002).

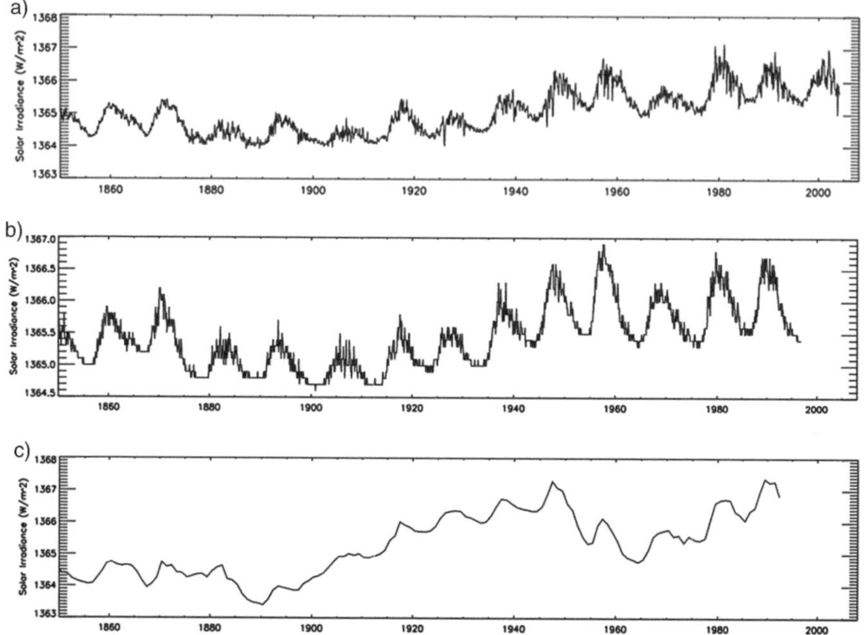

Fig. 6.3 Time series of different TSI reconstructions: (**a**) Solanki and Krivova, (**b**) Foster and (**c**) Hoyt and Schatten (Roy 2010)

Next Part

As oceans constitute three-fourth of the whole globe, the importance of ocean on our climate is univocal. The next part (Part II) initially focuses on the ocean and discusses various relevant definitions. It will then define solar so-called Top-Down and Bottom-up mechanism and will discuss a few related researches. Next, a general overview of solar influence on climate in a form flow chart will be presented. It discusses initiated by solar 11-year cyclic variability how the overall atmosphere–ocean coupling takes place. Those also involve QBO, ENSO, polar annular modes and climate change. Second chapter will mainly show why it is so difficult to detect robust solar influence on the climate of the Earth and indicates about complications.

References

Baliunas S, Jastrow R (1990) Evidence for long-term brightness changes of solar-type stars. Nature 348:520–522

Beer J et al (2000) The role of the sun in climate forcing. Quaternary Sci Rev 19(1-5):403–415

Brönnimann S, Compo GP, Sardeshmukh PD, Jenne R, Sterin A (2005) New approaches for extending the twentieth century climate record. Eos, Trans AGU 86(1):2

Foster S (2004) Reconstruction of solar irradiance variations, for use in studies of global climate change: application of recent SoHO observations with historic data from the Greenwich Observatory, PhD thesis, University of Southampton, Southampton, U.K, 2004

Gray LJ, Haigh JD, Harrison RG (2005) The influence of solar changes on the Earth's climate, Hadley Centre technical note 62, Met Office, Exeter. http://www.metoffice.gov.uk/publications/HCTN/HCTN_62.pdf

Hoyt DV, Schatten KH (1993) A discussion of plausible solar irradiance variations, 1700–1992. J Geophys Res 98(A11):18895–18906

Hoyt DV, Schatten KH (1998) Group sunspot numbers: a new solar activity reconstruction. Sol Phys 181(2):491

Lean J (2000) Evolution of the Sun's spectral irradiance since the maunder minimum. Geophys Res Lett 27(16):2425–2428

Lean J, Rind D (1998) Climate forcing by changing solar radiation. J Climate 11:3069–3094. https://doi.org/10.1175/1520-0442

Lean J, Beer J, Bradley R (1995) Reconstruction of solar irradiance since 1610: implications for climate change. Geophys Res Lett 22:3195–3198

Lean J et al (2005) SORCE contributions to new understanding of global change and solar variability. Solar Physics 230:27–53

Lockwood M, Stamper R, Wild MN (1999) A doubling of the sun's coronal magnetic field during the last 100 years. Nature 399:437–439. https://doi.org/10.1038/20867

Mann ME, Cane MA, Zebiak SE, Clement A (2005) Volcanic and solar forcing of the tropical Pacific over the past 1000 years. J Climate 18(3):447–456

Meehl GA (2002) The tropospheric biennial oscillation and Asian-Australian monsoon rainfall. J Climate 15:722–744

Reid G (1997) Solar forcing of global climate change since the mid-17th century. Clim Change 37:391–405

Roy I (2010) Solar signals in sea level pressure and sea surface temperature. Department of Space and Atmospheric Science, PhD Thesis, Imperial College, London

Smith RL et al (2003) A bivariate time series approach to anthropogenic trend detection in hemispheric mean temperatures. J Climate 16:1228–1240

Sofia S, Li LH (2001) Solar variability and climate. J Geophys Res 106:12969–12974

Solanki SK, Fligge M (1998) Solar irradiance since 1874 revisited. Geophys Res Lett 25(3):341–344

Solanki SK, Krivova NA (2003) Can solar variability explain global warming since 1970? J Geophys Res Space Physics 108(A5)

Stott PA, Tett SFB, Jones GS, Allen MR, Ingram WJ, Mitchell JFB (2001) Attribution of twentieth century temperature change to natural and anthropogenic causes. Climate Dynam 17:1–21

Wang Y-M, Lean JL, Sheeley NR Jr (2005) Modeling the Sun's magnetic field and irradiance since 1713. Astrophys J 625(1):522–538

Willson RC, Mordinov AV (2003) Secular total solar irradiance trend during solar cycles 21–23. Geophys Res Lett 30(5):1199–1202. https://doi.org/10.1029/2002GL016038

Part II
Atmosphere-Ocean Coupling and Solar Variability

Abstract In Part II, we mainly cover about Atmosphere and Ocean Coupling and the role of the sun in regulating the coupling. The second chapter is structured as follows. Chapter 7 mostly focuses the ocean and various relevant definitions. Chapter 8 talks about some contradictions related to the Sun and ENSO connection and how those can be possibly reconciled. It gives a critical insight about analysing scientific results. In Chap. 9, a debate is initiated about a combined influence of the Sun and ENSO. It shows how an apparent different/conflicting result can be resolved that lead forward to scientific understanding. Chapter 10 defines solar so-called 'Top-Down' and 'Bottom-up' mechanism and discusses related research. Finally, Chap. 11 presents a general overview of solar influence on climate in a form flow chart. It discusses initiated by solar eleven-year cyclic variability how the overall atmosphere-ocean coupling takes place. It also involves QBO, ENSO, polar annular modes and climate change. It shows why it is so difficult to detect robust solar influence on the climate of the Earth and indicates about complications.

Keywords Atmosphere-Ocean Coupling · Solar 'Top-Down' influence · 'Bottom-Up' influence · QBO · ENSO · Shallow meridional overturning circulation

Chapter 7
Ocean Coupling

Abstract This chapter initially focuses on the ocean and discusses various relevant definitions. Section 7.1 discusses shallow overturning circulation, ocean conveyor belt and global wind-driven ocean circulation. Section 7.2 describes ENSO, and focus is on various issues: (i) ENSO, thermocline and upper ocean heat content, (ii) ENSO and delayed oscillator theory, (iii) ENSO and shallow MOC in tropical Pacific and (iv) pycnocline convergence vs. Nino region sea surface temperature (SST).

Keywords ENSO · Thermocline · Shallow overturning circulation · Ocean Conveyor Belt · Ocean heat content · Thermohaline circulation

7.1 Shallow Overturning Circulation

There are two major systems of ocean current: the first one is the surface current which is a wind-driven, shallow and fast moving; the second one is the deep current driven by density differences – which is mostly deep, slow moving and massive.

Throughout most of the ocean, layers form three principal zones: The surface zone (also known as the mixed layer), the pycnocline zone (pycno means density, and cline is slope) and the deep zone. The surface zone is usually confined within a layer from the surface ocean to a depth of about 100 m. The pycnocline separates the upper water mass and surface layer from the deep water mass and is often termed as the thermocline (as temperature also changes rapidly with depth). Since temperature (and density) decreases (increases) rapidly with depth, this zone is highly stable; mixing is suppressed, and the upper and lower current system acts independently. Wind acting on the surface of the ocean causes a partial transfer of kinetic energy from the wind to water. Wind-driven currents decline with depth and are limited by the permanent pycnocline – around 100–200 m of depth (but in some cases, even as deep as 1000 m). The global pattern of wind causes major ocean currents in the surface layer. Deep and shallow waters respond to different forces and have different circulation pattern. These ocean features are also discussed in Roy (2010).

© Springer International Publishing AG, part of Springer Nature 2018

I. Roy, *Climate Variability and Sunspot Activity*, Springer Atmospheric Sciences, https://doi.org/10.1007/978-3-319-77107-6_7

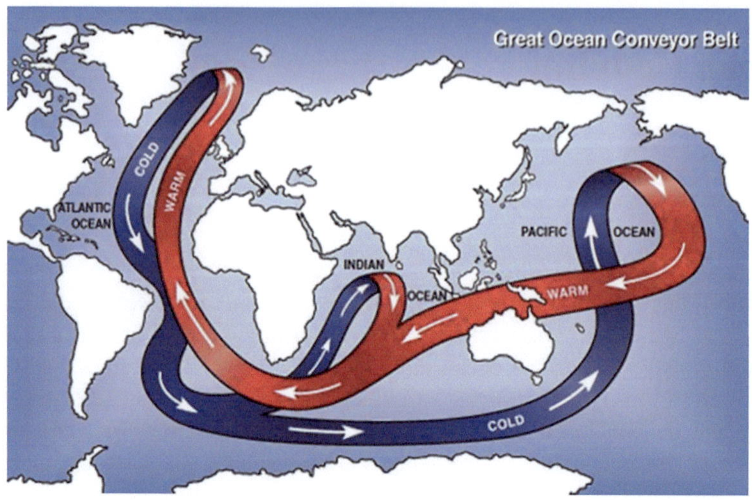

Fig. 7.1 Ocean conveyor belt. (Source: https://climatekids.nasa.gov/ocean/, image accessed on 02/04/2018)

Solar energy governs the circulation of the atmosphere and thus the winds that drive the ocean currents in the surface layer. Atmospheric circulation has a period ranging from days/months to a ~year; for ocean surface current, it is weeks/months, and for deep ocean circulation, it is 500–3500 years (http://instaar.colorado.edu/~lehmans//env-issues/documents/S09_3520_10comp.pdf). Thus, the deep ocean circulation only plays a major role in a scale of longer time climate.

The circulation of ocean mainly operates through a global scale conveyor belt shown in Fig. 7.1. Surface water around the north Pacific, in the vicinity of the Aleutian Low (AL), is warmed up by the Sun and moves westward across the Indian and Pacific Ocean. Crossing the equator, it reaches north of Atlantic and loses much of its heat. Being cold, it then sinks and afterwards proceeds to the Pacific via the Antarctic through a deep ocean current. The ocean conveyor belt is also known as thermohaline circulation as temperature (hence thermo) and salinity (hence haline) both drives the conveyor. In Fig. 7.1, the warm shallow current is shown by red colour, and the cold deep ocean circulation is marked by blue.

On occasion, the thermohaline circulation is referred to as the meridional overturning circulation which is often abbreviated as MOC. It is however, more accurate, because it's difficult to segregate the part of the circulation driven by salinity and temperature alone, as opposed to other important factors, e.g. winds. The shallow MOC is characterised by equatorward geostrophic volume transport convergence across the ocean interior of pycnocline at 9°N and 9°S. It connects the regions of subtropical subduction to tropical pycnocline. It is one component of the subtropical cells (STCs) as discussed by McCreary and Lu (1994). Over ocean basins, subduction occurs predominantly through the gyre-scale circulation. Within a basin, the surface wind stress drives cyclonic and anticyclonic recirculations, which are often referred as subpolar and subtropical gyres, respectively (Fig. 7.2).

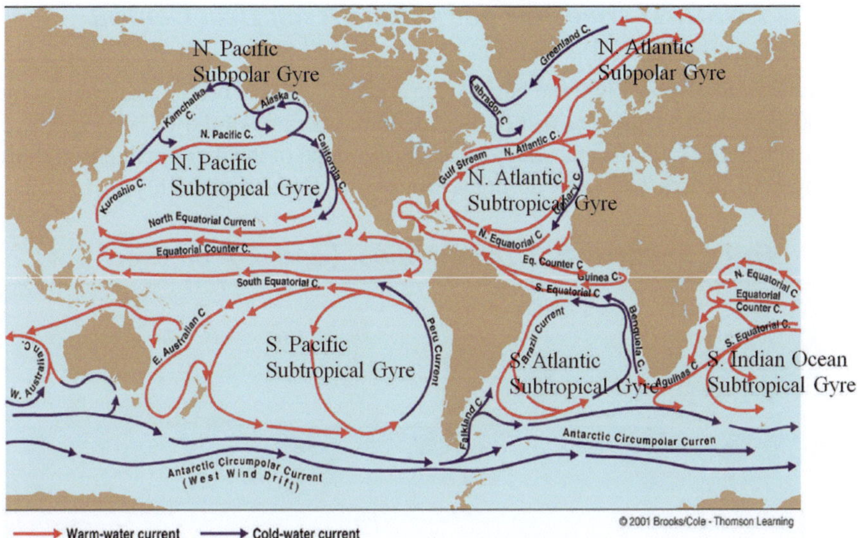

Fig. 7.2 Global wind-driven ocean circulation. (Source: http://www4.ncsu.edu/~ceknowle/ Envisions/chapter09copy/part1.html, image accessed on 2/4/2018)

The downwelling of surface fluid over the subtropical gyre is induced by the wind forcing.

In what follows, we will mainly focus on the shallow MOC in the Pacific region, which may act as a bridge between tropical and extratropical (north) mass exchange. At least two different mechanisms have been proposed for such a connection (Yu and Rienecker 1999). The first one emphasises the existence of shallow ocean pathway that allows the north Pacific to influence ENSO through ocean subduction. The other one argues that decadal-scale variations in the atmospheric circulation over the northern Pacific can extend to the tropics and build that connection. However, it states that the slow variations in the trade wind system precondition the mean state of the thermocline in the equatorial Pacific. It is observed that ENSO events regularly disrupt the Pacific surface flows, and hence it is necessary to know more about the ENSO to understand SST in the Pacific.

7.2 ENSO

In Part 1, having defined ENSO, here the mechanism part, some of its characteristics and its relevance to the tropical–extratropical mass exchange in the Pacific, through the shallow MOC will be discussed.

7.2.1 ENSO, Thermocline and Upper Ocean Heat Content

In Fig. 7.3, it is shown that the basin-wide equatorial upper ocean heat content (at 0–300 m) is highest prior to and during the initial periods of the El Niño and vice versa during the La Niña episode. Moreover, the slope of the thermocline is highest during cold phases and least during warm phase. Here the monthly thermocline

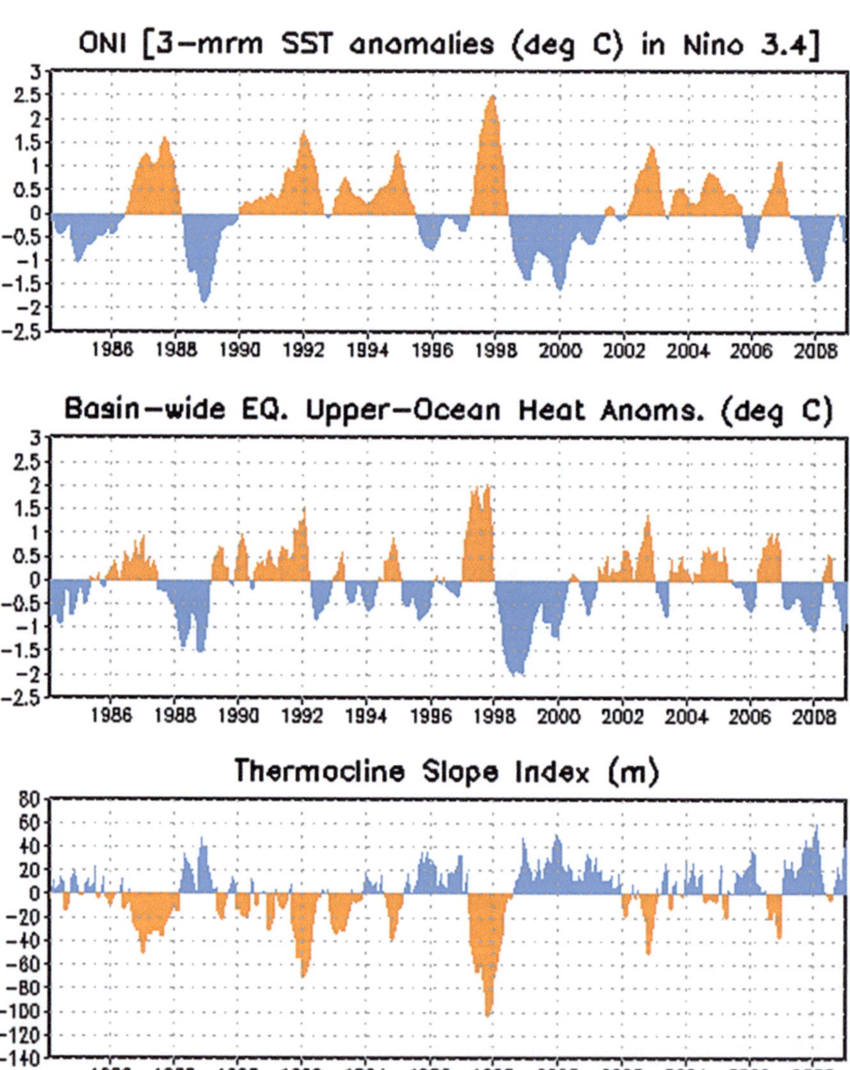

Fig. 7.3 Upper ocean conditions in the equatorial Pacific: (**a**) oceanic Niño 3.4 index, (**b**) upper ocean heat anomaly and (**c**) thermocline slope index. Warm episodes (solid red lines) and cold episodes (purple dashed line) are marked. (Source: http://www.cpc.noaa.gov/products/analysis_monitoring/lanina/enso_evolution-status-fcsts-web.pdf, link accessed on 2/4/18)

slope index has been defined as the difference in anomalous depth of the 20° isotherm between the eastern (90° to 140°W) and western Pacific (160°E to 150°W). Hence, the upper ocean heat content is sometimes considered as a precursor of the ENSO formation.

For the generation of ENSO, there needs to be a phase-transition mechanism able to provide a negative feedback, to reverse its cycle and account for the period associated with the cycle. The delayed oscillator theory (Cane et al. 1990; Battisti and Hirst 1989) is one of the most well-accepted theories, which can successfully explain ENSO behaviour.

7.2.2   ENSO and Delayed Oscillator Theory

The delayed oscillator suggests that the phase-transition mechanism for the cycle of ENSO (i.e. memory) is provided by oceanic Rossby and Kelvin waves. It is forced by atmospheric wind stress around the central Pacific. Its period is determined by the propagation and reflection of waves, along with local sea–air coupling.

Such a phase transition of the ENSO cycle is shown in Fig. 7.4a–d for a warm event of the ENSO, though the movement of the thermocline is not captured in that figure. Wind forcing at the central Pacific generates a downwelling Kelvin wave (K) which propagates eastward, alongside an upwelling Rossby wave (R), that propagates westward (Fig. 7.4a). Figure 7.4b shows the first Kelvin wave reaches the

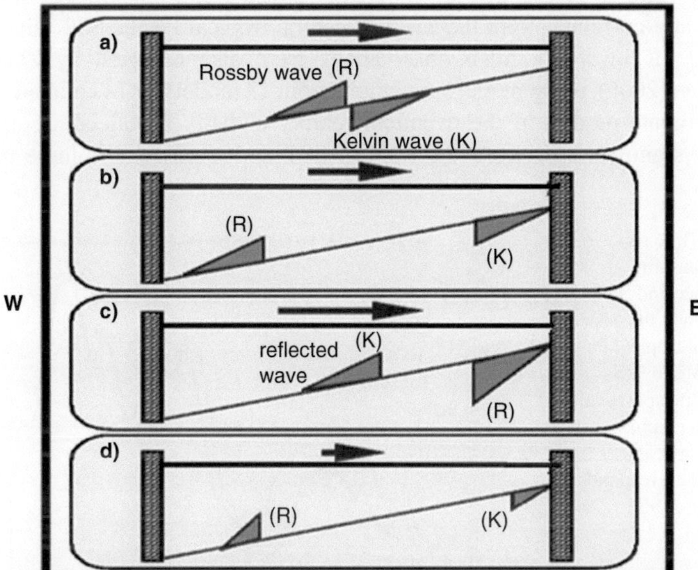

Fig. 7.4 Phase transition of oceanic Kelvin and Rossby wave. (Source: after http://www.ess.uci.edu/~yu/class/ess200a/lecture.7.climate.variatons.pdf, link accessed on 2/4/18)

eastern basin (shown with E) and causes shifting down of the thermocline there (not shown in that figure). It subsequently allows warm water of the western basin (shown with W) to move eastward causing the warm event of ENSO the El. Niño. The slow Rossby wave is reflected at the boundary of the west, at a later time. It then reverts as an upwelling Kelvin wave and propagates towards the eastern basin causing uplifting the thermocline (not shown) and thus reversing the phase of ENSO. Likewise, the Kelvin wave being reflected in the eastern basin (E) also relapses as a downwelling slow-moving Rossby wave (Fig. 7.4c).

Finally, the westward moving Rossby wave after reflection again hits the east tropical basin as a downwelling Kelvin wave, deepening the thermocline again and causing little warming at the eastern Pacific coast (Fig. 7.4d). The period of ENSO is determined by the propagation time of these two waves. Following a similar mechanism, the wind forcing at the central Pacific in a reverse direction can incite the cold event of ENSO.

The delayed oscillator theory of ENSO suggests that the basin of the ocean needs to be wide enough to generate the delaying by ocean wave propagation and reflection. It is satisfied by the Pacific Ocean. It is believed that the Atlantic Ocean could generate ENSO-like oscillatory behaviour if the external forcing is applied. But the Indian Ocean is too small to establish ENSO-like behaviour.

7.2.3   ENSO and Shallow MOC in Tropical Pacific

The observational results of Fig. 7.5 clearly indicate that the strength of shallow MOC is anti-correlated with the east and central tropical Pacific SST (since transport scale is inverted). Such observations were also explored by Zhang and McPhaden (2006), using model simulations from 18 models by 14 coupled state-of-the-art climate models of the twentieth-century climate. It noticed a correlation, which is significant, between the tropical SST and meridional volume transport

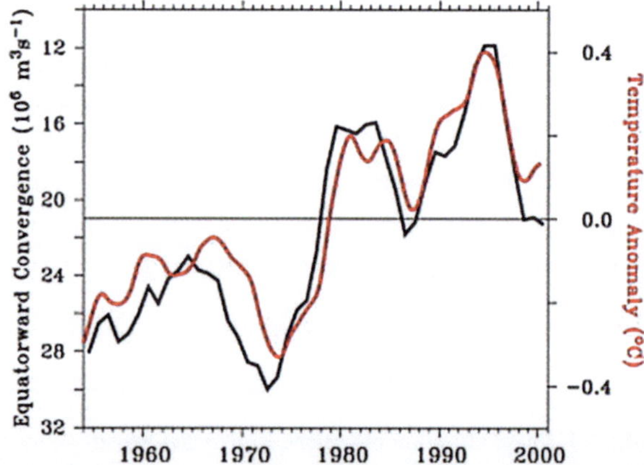

Fig. 7.5 Time series of volume transport convergence and SST averaged over the eastern and central tropical Pacific, 90–180°W and 9°S–9°N. The transport scale is inverted to emphasise that a slowdown of the circulation corresponds to a warming of SST (Zhang et al. 2006)

convergence over the last half century, and it is detected in a majority of the models. It suggests that as well as the change in thermocline slope in the eastern Pacific and the heat content of the upper ocean around the equatorial Pacific basins, ENSO cold phase is also linked with the strengthening of the shallow overturning circulation.

7.2.4 Pycnocline Convergence vs. SST

Studies (McPhaden and Zhang 2002 among others) indicate that, during the 1950s to 1990s, a decline in the shallow overturning circulation is noticed in the tropics of Pacific (also shown in Fig. 7.6), with a rebound since 1998 (McPhaden and Zhang 2004). An observational study suggests that during 1998–2003 the convergence of cold ocean interior water of pycnocline towards the equator has risen to 24.1 ± 1.8 x 106 m^3 s^{-1} (McPhaden and Zhang 2004, see Fig. 7.6) from a low of 13.4 ± 1.6 x106 m^3 s^{-1} during 1992–1998. There is a trend noticed towards reduced transport convergence of 11 Sv (1 Sv = $10^6 m^3 s^{-1}$), over the period 1953–2001 (Zhang et al. 2006). According to them, the circulation on decadal time scales fluctuates significantly, and the variations in decade-to-decade scale are of 8–12 Sv about the trend.

The time of the noted reduction in circulation is aligned with the changes in the strength of the Walker cell. As mentioned above and supported by Fig. 7.5, the ENSO and the shallow overturning circulation are connected via an ocean pathway so that a climate change signal in both ENSO and the Walker circulation is also likely to be detected in the shallow MOC in the Pacific.

Different forces drive the tropical Hadley circulation and Walker circulation. For the Hadley circulation, meridional differential heating of the global radiative pro-

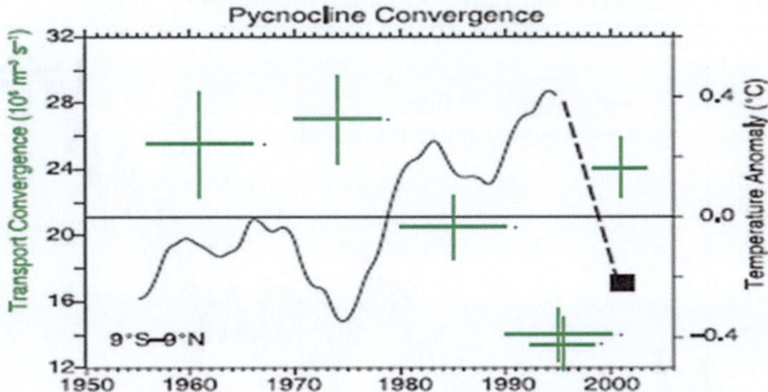

Fig. 7.6 Equatorial SST anomalies (black) for 1950–2003, and meridional volume transports in the pycnocline across 9°N and 9°S (with green). It is calculated where equatorial upwelling is most prevalent (across same latitude belt and 90°W–180°W). Anomalies are relative to 1950–1999 averages. The temperature time series is smoothed using a 5-year running mean (twice) to filter out year-to-year oscillations and the seasonal cycle associated with ENSO. The single 5-year average (July 1998–June 2003) is connected to the smoothed series using a dashed line (McPhaden and Zhang 2004)

cess is the essential driving force, and hence it should exist even under a hypothetical aqua-planet without topography or sea–land contrast. Whereas to generate/alter the Walker circulation, land–ocean distribution plays a paramount role. The Walker cell tends to weaken in models, when it is forced with increase in amount of greenhouse gases (Meehl et al. 2007; Vecchi et al. 2006). The Hadley circulation also weakens but less than the Walker circulation (Vecchi and Soden 2007). Key observations, relating to tropical circulations and shallow MOC in the tropical Pacific, are highlighted in the box below as they are important for ocean–atmosphere coupling and will frequently be referred in subsequent discussions.

- Strong decrease in strength of shallow MOC around tropical Pacific after the 1950s
- Modest intensification since 1998
- Also true for Hadley and Walker circulation; more in the Walker than Hadley circulation (McPhaden and Zhang 2004; Vecchi and Soden 2007)

7.2.5 Abrupt Rise in Temperature During 1977–1998

Figure 7.7 indicates how much warmer or colder the world was than the average temperature during 1951–1980. The uncertainties in this plot are shown by the green bars. As seen in Fig. 7.7, there was an abrupt rise in temperature during 1977–1998. Since 1998, the global land–ocean temperature index is not showing a similar rising trend.

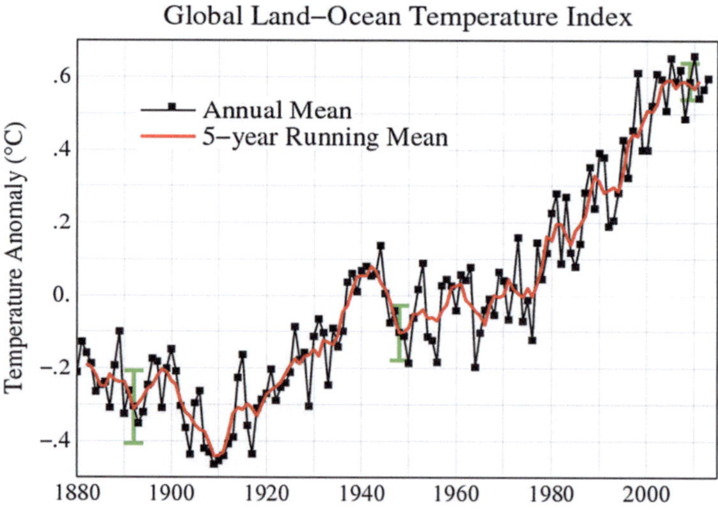

Fig. 7.7 Global mean surface air temperature, from the NASA Goddard Institute for Space Studies. (Source: http://data.giss.nasa.gov/gistemp/)

References

Battisti DS, Hirst AC (1989) Interannual variability in a tropical atmosphere–ocean model: influence of the basic state, ocean geometry and nonlinearity. J Atmos Sci 46:1687–1712

Cane MA, Muennich M, Zebiak SE (1990) A study of self-excited oscillations of the tropical ocean–atmosphere system. Part I: linear analysis. J Atmos Sci 47:1562–1577

McCreary J, Lu P (1994) On the interaction between the subtropical and the equatorial oceans: the subtropical cell. J Phys Oceanogr 24:466–497

McPhaden MJ, Zhang D (2002) Slowdown of the meridional overturning circulation in the upper Pacific Ocean. Nature 415(6872):603–608

McPhaden MJ, Zhang D (2004) Pacific Ocean circulation rebounds. Geophys Res Lett 31:L18301. https://doi.org/10.1029/2004GL020727

Meehl GA et al (2007) Global climate projections. In: Solomon S et al (eds) Climate change 2007: the scientific basis: contribution of working group I to the fourth assessment report of the Intergovernmental Panel on Climate Change. Cambridge University Press, Cambridge, pp 747–845

Roy I (2010) Solar signals in sea level pressure and sea surface temperature. Department of Space and Atmospheric Science, PhD Thesis, Imperial College, London

Vecchi GA, Soden BJ (2007) Global warming and the wakening of the tropical circulation. J Climate 20:4316–4340

Vecchi GA, Soden BJ, Wittenberg AT, Held IM, Leetmaa A, Harrison MJ (2006) Weakening of tropical Pacific atmospheric circulation due to anthropogenic forcing. Nature 441. https://doi.org/10.1038/nature04744

Yu LS, Rienecker MM (1999) Mechanisms for the Indian Ocean warming during the 1997–98 El Niño. Geophys Res Lett 26:735–738

Zhang D, McPhaden MJ (2006) Decadal variability of the shallow Pacific meridional overturning circulation: Relation to tropical sea surface temperatures in observations and climate change models. Ocean Model 15(3–4):250–273

Chapter 8
The Sun and ENSO Connection–Contradictions and Reconciliations

Abstract This chapter focuses on solar signal and ENSO connection (if any) in recent centuries. The discussion begins with two contradictory findings: **Contradiction (I),** solar signal on tropical Pacific SST-active solar years and ENSO and **Contradiction (II),** solar signal on tropical Pacific SST – El Niño or La Niña. It reconciled those contradictions and additionally proposed two different mechanisms. The mechanisms differ in the earlier period to that from the later.

Keywords ENSO · Solar signal · Multiple Linear Regression · Compositing technique · Warm event of ENSO · Cold event of ENSO · Shallow meridional overturning circulation

8.1   Solar Signal and ENSO

A quasi-decadal underlying variability is detected in the interannual ENSO by many researches (Chen et al. 2004; White and Liu 2008; Zhang et al. 1997). It questions whether the Sun is having any effect on the ENSO. Roy et al. (2016) discussed that the region of Aleutian Low is influenced by both the sun and the ENSO, though that feature is seasonal dependent and only present during DJF. After adding an 11-year-period cosine signal to the solar constant (of amplitude ~2.0 W m^{-2}) in the fully coupled atmosphere–ocean GCM, White and Liu (2008) could simulate both the ENSO and the quasi-decadal oscillation (QDO). The GCM used is the Fast Ocean–Atmosphere Model (FOAM) of Jacob et al. (2001). Though they used the term 'QDO', but it is similar as DSO. Their model QDO was shown similar in patterns and evolution with the observed one. On the other hand, in the absence of solar 11-year cycle, the model only simulates the ENSO.

Using the method of solar maximum compositing, van Loon et al. (2007) (hereafter, vL07) detected a very strong tropical solar signal, resembling that of the negative phase of ENSO (as shown and discussed in Fig. 5.6 in Part I, Chap. 5). Moreover, in the midlatitude of Pacific, they detected a significant positive solar fingerprint. The method of solar maximum compositing is described in Part I (Chap. 5), with schematic (Fig. 5.7). Using the same dataset over the same period as

© Springer International Publishing AG, part of Springer Nature 2018
I. Roy, *Climate Variability and Sunspot Activity*, Springer Atmospheric
Sciences, https://doi.org/10.1007/978-3-319-77107-6_8

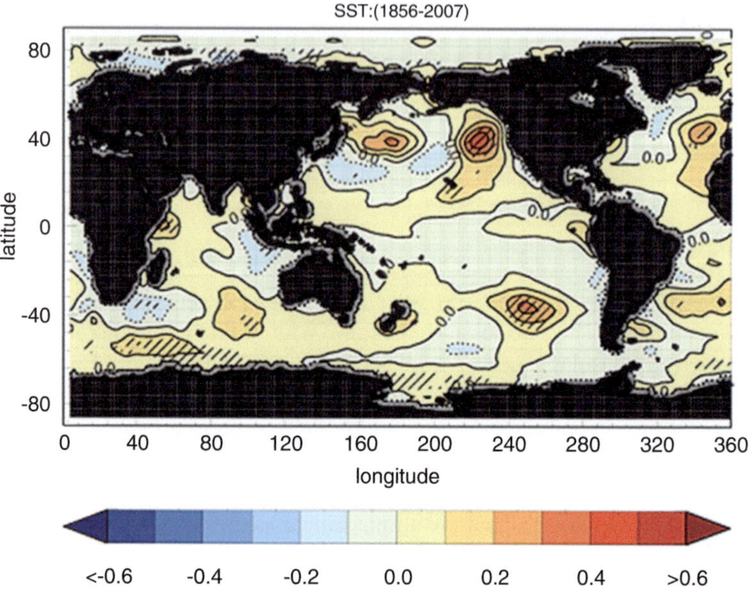

Fig. 8.1 Amplitudes of the components of the SST (using dataset from NOAA) variability (in °K), due to the Sun. Other independent factors used are OD, trend and ENSO. All months of a year are used which is represented by monthly mean values, with annual cycle removed (Roy 2014)

used by vL07, Roy and Haigh (2010) could not detect −ve solar signal around tropical Pacific using multiple linear regression (MLR) technique, as shown in Fig. 8.1. The method of MLR technique is discussed in details in Part I, Chap. 5 with relevant figure (Fig. 5.2). In Fig. 8.1, NOAA dataset is used, but even using another dataset HadSST2, from Hadley Centre, MLR technique could not detect strong negative signal around tropical Pacific as observed by vL07.

To find the discrepancy between these two studies, a scatter plot is presented (Fig. 8.2, Top) with ENSO during DJF vs. sunspot number (SSN). No apparent relationship is seen, and this is also verified by a separate regression analysis (not presented here). But when, maximum years of solar cycles are marked in Fig. 8.2 (bottom), 9 out of 14 have a ENSO value less than the mean, moreover, four (2000, 1989, 1917, 1893) are linked with particularly strong cold ENSO phase. It is only the solar maximum/peak years as used by vL07 to represent the solar signal; it is evident that the result will match a La Niña or cold event-like situation. Thus using the method of compositing of solar max, the signal of ENSO may be misinterpreted as solar. Hence, it remains to be determined to what extent the observed signature can be assigned as due to solar variability instead of ENSO.

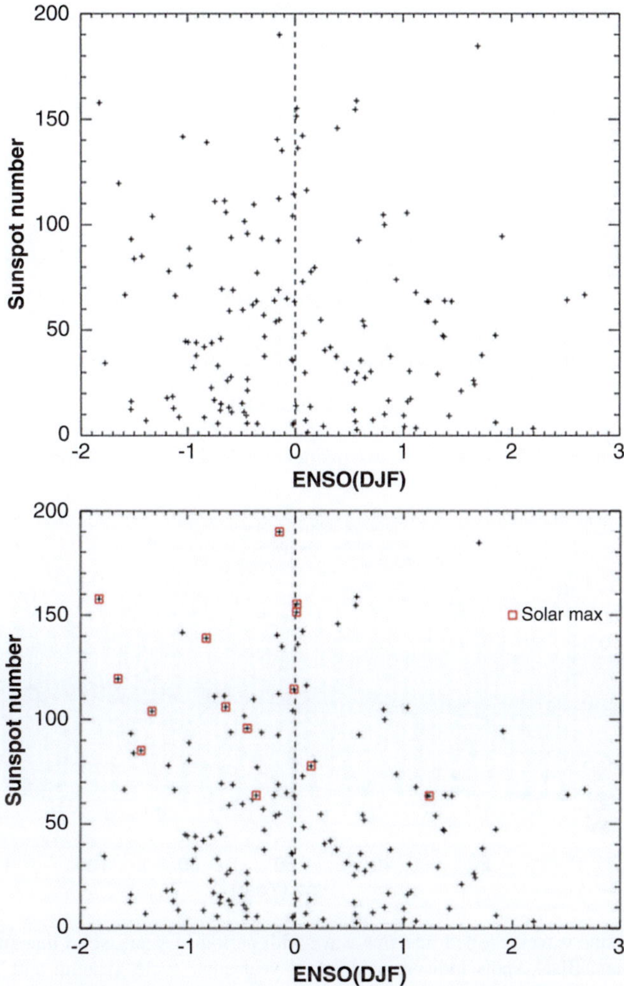

Fig. 8.2 Scatter diagram of annual mean SSN vs. DJF means ENSO index during (1856–2007) (top), same as top but with solar maximum years identified (bottom) (Roy 2010)

8.2   Contradiction (I): Solar Signal on Tropical Pacific SST-Active Solar Years and ENSO

Figure 8.3 is a similar diagram as Fig. 8.2 (bottom) of annual mean SSN vs. DJF mean ENSO index with solar maximum years identified. But here two quarters for SSN above 80 are marked. It is noticed that all solar max years (marked by red) with high SS number (>80) are on cold event side. But it is not always true for all years with high SS values (shown by right quarter).

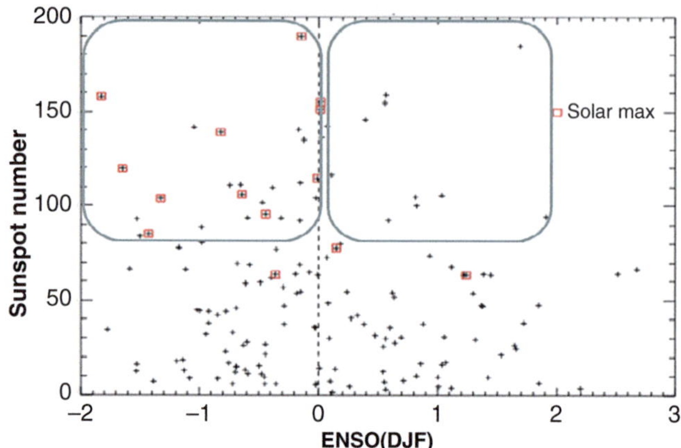

Fig. 8.3 Same as Fig. 8.2 (bottom), but two quarters for SSN above 80 are marked

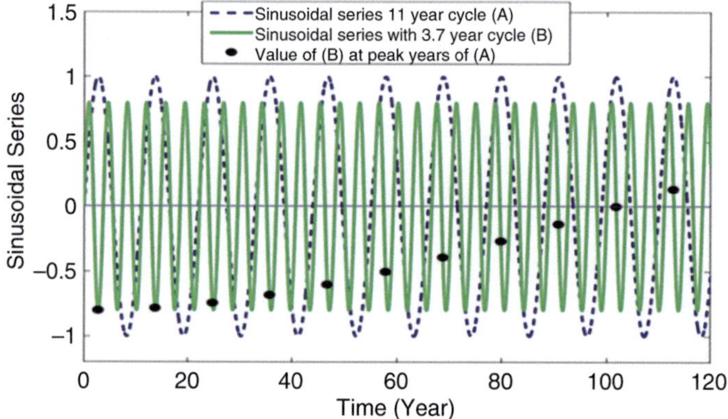

Fig. 8.4 Two sine waves: dashed line sine wave with period 11 years, solid line sine wave with period 3.7 years. Black spots indicate peak of 11-year solar cycle. (Hamm and Tung (2012), ©American Meteorological Society. Used with permission)

Hamm and Tung (2012) proposed it could be a mere coincidence. Using two sine waves in Fig. 8.4, they showed it is possible that ENSO cold event situation can coincide with the peak year of solar cycle for many solar cycles. In that figure, two sine waves are shown, one with the dashed line representing sine wave with period 11 years, the other by a solid line with period 3.7 years. Black spots indicate peak years of 11-year cycle matches with the negative phase of 3.7 years cycle for several 11-year cycles.

Figure 8.5 is the scatter diagram of annual mean SSN vs. DJF mean ENSO index during two different periods, one (a) (1856–1957) and (1998–2007) and the other (b) (1958–1997). It is seen, before 1958 and also since after 1998 (Fig. 8.5a), all years with higher SSN (say, above 80) are with –ve ENSO index. Possibly, active Sun (peak as well as high solar years) influences SSTs, but this is overwhelmed by

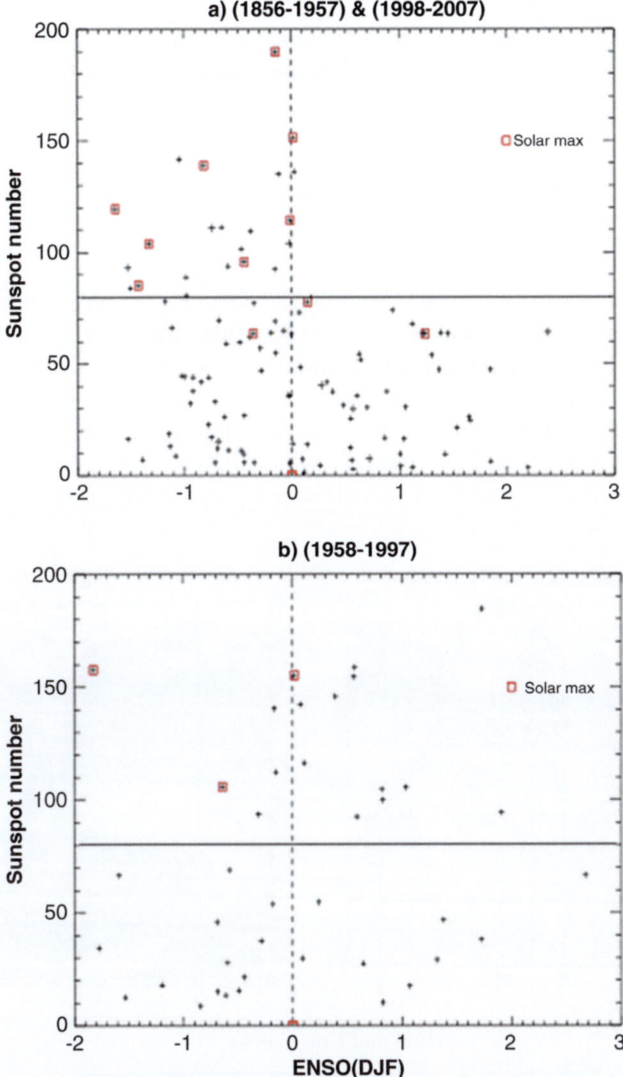

Fig. 8.5 Scatter diagram of annual mean SSN vs. DJF means ENSO index during: (**a**) (1856–1957) and (1998–2007) and (**b**) (1958–1997) (Roy 2014)

inherent strong variability of ENSO at the lower activity period of the Sun. Thus, Sun ENSO connection was different before the 1950s (and after 1997), and this applies to all years irrespective of peak or high Sun. On the other hand, such a bias for HS years is missing during 1958 to 1997 (Fig. 8.5b). The ENSO-related signal is confined in the Pacific for high solar years is also shown by Kodera et al. (2007) and according to them it is mainly originated from a shift in the position of the descending branch of anomalous Walker cell. Following the observation (Vecchi and Soden 2007; McPhaden and Zhang 2004), as highlighted with a box (Sect. 7.2.4), it could change in ocean and atmosphere circulation during that period. Those could be possible causes for the modified solar–ENSO behaviour.

8.3   Contradiction (II): Solar Signal on Tropical Pacific SST-El Niño or La Niña

Figure 8.6. shows the solar signal on tropical Pacific SST as observed by three different studies and also discussed in Chap. 5. White et al. (1997) (hereafter, W97) using EOF (empirical orthogonal function) analysis for the period 1950–2000, detected warming pattern matching with 11-year solar cycle (also shown in Fig. 5.4). vL07 using the method of solar maximum compositing detected cooling signature (also shown in Fig. 5.6). Whereas White and Liu (2008) (hereafter, WL08), using the method of composite and SVD analysis of nine solar cycles for the period

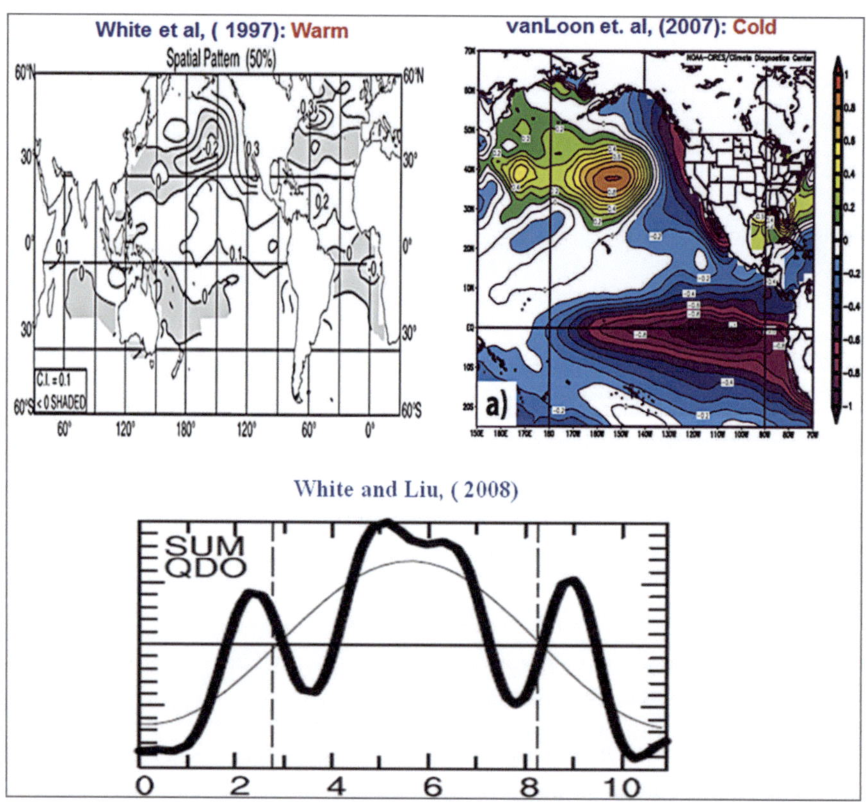

Fig. 8.6 Solar signal on tropical Pacific SST as detected by three different studies: White et al. (1997), vL07 and White and Liu (2008)

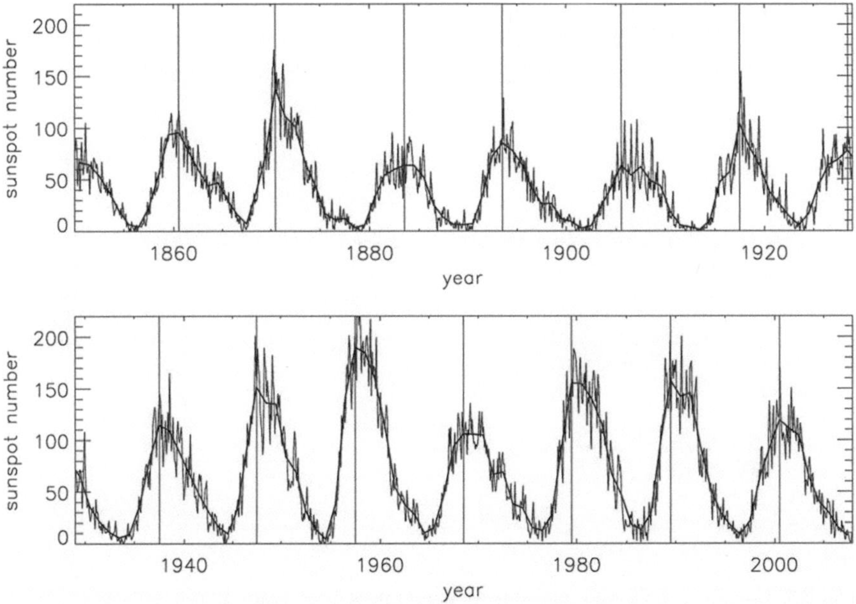

Fig. 8.7 Monthly mean SSN (thin line) and the annual mean (thick line). Vertical lines show the peak year of annual SSN for each 11-year cycle (Roy and Haigh 2010)

1900–2000, observed a phase locking with solar 11-year cycle (also shown in Fig. 5.5). They found warm event (WE) signature near the peak of smoothed decadal solar oscillation (DSO) is typically preceded and followed with a lag/lead of about 2 years, by a cold event (CE). Here we would address what are the main reasons for all these contradictions.

Figure 8.7 indicates the time series of monthly SSNs. It also marks the years of highest annual SSN, to identify peak solar years (as used by vL07). It is apparent that peak solar years tend to occur very soon after the cycle is active and seen for all the stronger cycles. It occurs at least a year in advance from the peak of a broader decadal variation. Changes in TSI however tend to follow the latter (e.g. Krivova et al. 2003; Lean, 2000). Hence, the analysis that follows peak SSN years represents the solar signal at a particular phase of the solar cycle (such as the vL07 results). Whereas the works of W97 represent more broadly the difference between periods of lower and higher SSN. It explains why observations of W97 differ from vL07 – as years of annual peak SSN (used by vL07) falls a year or more in advance of a maximum of smoothed decadal solar oscillation (DSO), and usually it takes about 2 years for the ENSO to alter from cold phase to warm. The work of WL08 also provides an indication as to why a WE signature near the peak of the DSO is typically followed/preceded with a lag/lead of ~2 years, by a CE.

A peak in irradiance at the peak of the DSO indicates the La Niña event-like signature as shown by Meehl et al. (2008, 2009), which is lagged by an El Niño-like response after a year or two. Relating to a timing of this forcing and responses,

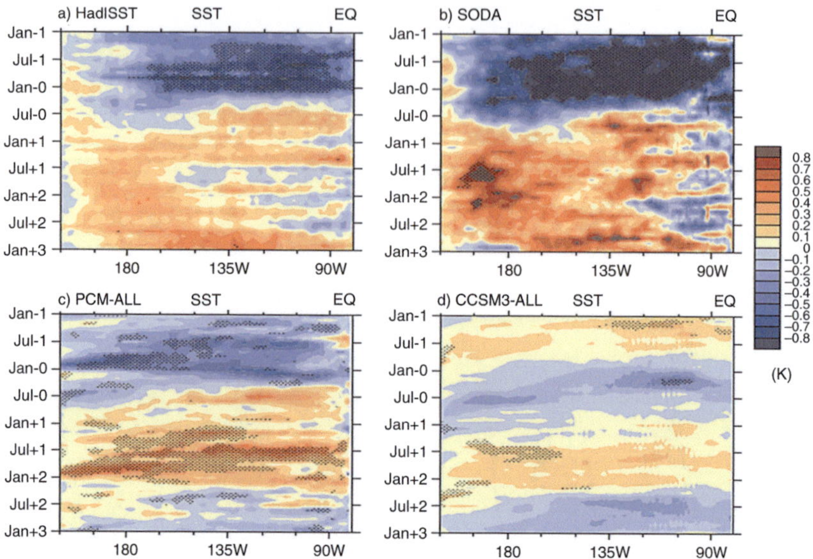

Fig. 8.8 Composite SST w.r.t. climatology during peak SSN years, shown as 'year 0', a year before shown as 'year -1' and 3 years following peak SSN years, shown as 'years +1, +2 and +3'. (Bottom) using PCM model (c) composite of four model simulations and using CCSM3 model (d) composite of five simulations. (Top) observations from HadISST data (**a**) and SODA ocean reanalysis (**b**). (From Meehl and Arblaster 2009. Reprinted with permission from AAAS; Roy 2014)

Meehl and Arblaster (2009) also presented results for SST composite during years of peak SSN, shown as 'year 0', a year before, shown as 'year 1' and 3 years following peak SSN years shown as 'years +1, +2 and +3'. They compared results of two models with observation and reanalysis product (as shown in Fig. 8.8). Among these two models, Parallel Climate Model (PCM) and Community Climate System Model version 3 (CCSM3), the former suggests closer fit to their two different observational results, one from HadISST and another from ocean reanalysis product, simple ocean data assimilation (SODA). These studies indicate that the timings of forcing and response are also important to understand SST solar connection. The discrepancies regarding tropical Pacific SST between two models mainly originate from different variability of ENSO within the models (Meehl and Arblaster 2009).

Based on the study of Fig. 8.8 and above discussions, Meehl et al. (2009) argued that all previous studies are in agreement, and all results of Fig. 8.6 are consistent. However, if all individual cycles are considered as is done in Table 8.1, it suggests differently.

Table 8.1 shows the value of ENSO (DJF) for the 14 solar cycles during 1856–2007, at the year of peak SSN and the following 2 years. The period of intensified tropical Pacific meridional overturning circulation (MOC) is shown by the horizontal lines between solar cycles 19 and 18 and between cycles 23 and 22. The alphabet 'W' and 'C' show that the index of ENSO-3.4 is higher or lower than 0.02 units, respectively (from its mean). A near neutral state is indicated by a dash.

Table 8.1 The ENSO (DJF) value at the year of peak SSN, and the following two years, during 1856–2007. 'W' denotes warm phase, 'C' cold phase and '–' neutral

Solar cycle no	Years	Peak year	State of ENSO (DJF)		
			peak year	1 y after peak y	2 y after peak y
10	1856-1867	1860	C	C	C
11	1867-1878	1870	C	C	C
12	1878-1890	1883	C	-	W
13	1890-1901	1893	C	C	C
14	1901-1913	1905	W	W	C
15	1913-1923	1917	C	C	W
16	1923-1933	1928	W	C	W
17	1934-1944	1937	-	C	C
18	1944-1954	1947	-	W	C
19	1955-1964	1957	C	W	W
20	1964-1976	1968	C	W	W
21	1976-1986	1979	-	W	C
22	1986-1996	1989	C	W	W
23	1996-2007	2000	C	C	C
Total			9 C 3 – 2 W	7 C 1 – 6 W	8 C 0 – 6 W

Roy and Haigh 2012, ©American Meteorological Society. Used with permission

Table 8.1 suggests of the nine cycles identified as cold at the peak SSN year; five turn warm after 1–2 years while four remain cold for the following 2 years. Thus, the 'lagged warm event' response as suggested by Meehl et al. (2008, 2009) is seen only for 5 cycles out of the 14. W97, that used data covering cycles 19–22, happen to include four 'lagged warm response' cases. These four cycles coincide with the period of weaker Walker and tropical oceanic meridional overturning circulation. During this time, ENSO activity was considerably higher too. Thus, no consistent ENSO-like variation is identified following peak years of the SSN cycle in tropical SSTs. In the available data, such behaviour happens only for 5 solar cycles out of 14 in total. The solar influence was overwhelmed by a stronger ENSO variability during the period 1958–1997. However, during the previous century, the more subtle effect of the Sun was better able to manifest itself.

Figure 8.9 shows results of multiple linear regression on SST during (1958–1997). Other independent factors are OD and trend for left one and trend, OD and ENSO for right. As is seen in the regression that the omission of ENSO gives a false signal of the sun- it suggests a warming (cooling) of an active (less active) Sun on SST in tropics during a later period. However during 1856–1957, the solar signal on SST in tropical Pacific is same with or without ENSO in regression and shown in Fig. 8.10 (Roy 2013).

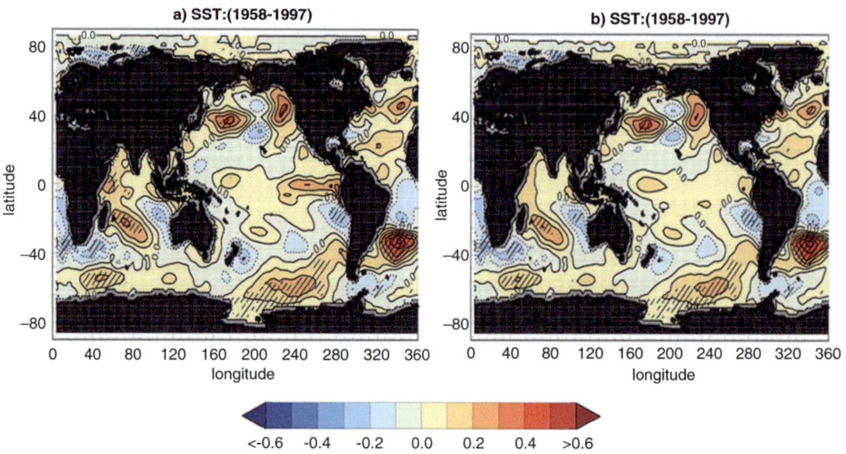

Fig. 8.9 Amplitudes of the solar components of the variability (using monthly SSN) of SST. Period considered is (1958–1997), while other independent factors are (**a**) optical depth (OD) and trend and (**b**) OD, trend and ENSO. After removing annual cycle, regression was done annually. (Roy 2014)

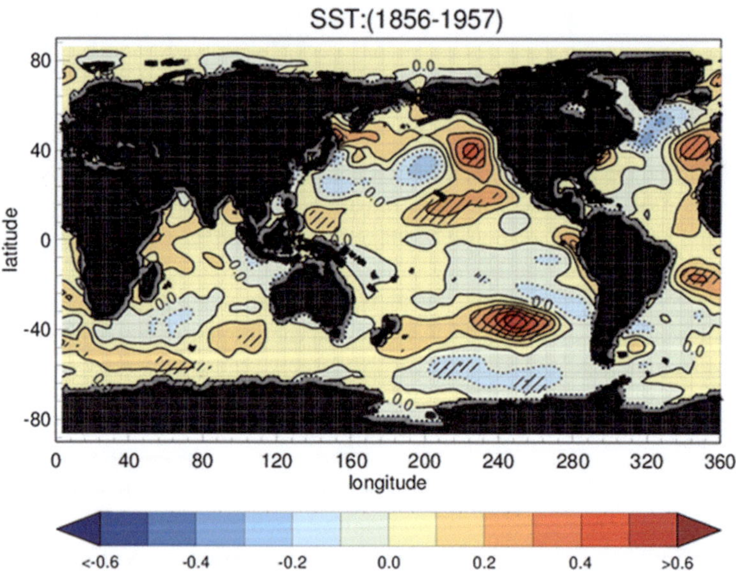

Fig. 8.10 Amplitudes of the solar components of the variability (using monthly SSN) of SST. Period considered is 1856–1957, and other independent factors are OD and trend. Including ENSO also, the signal is the same in the tropical Pacific. After removing annual cycle, regression was done annually

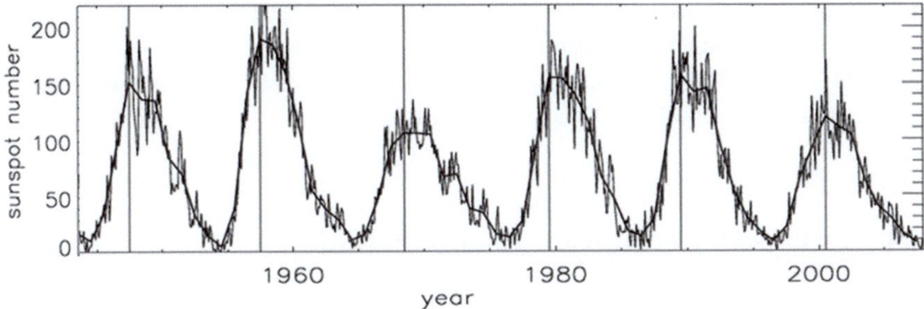

Fig. 8.11   Same as Fig. 8.7 but only during latter half of twentieth century. (After Roy and Haigh 2012)

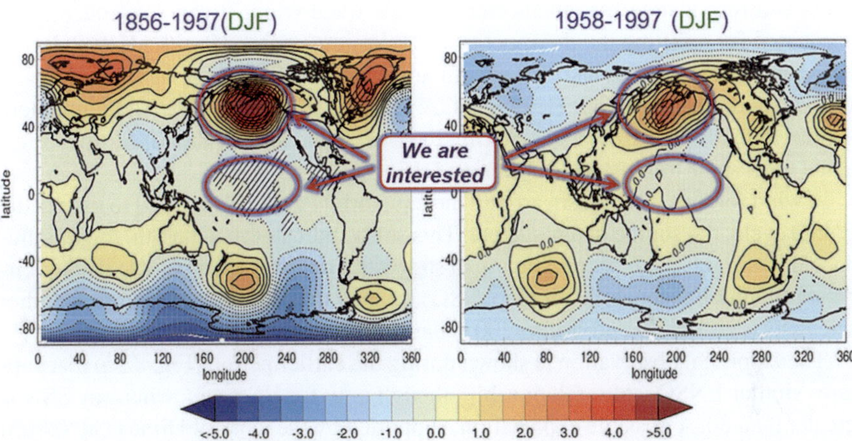

Fig. 8.12   Amplitudes of the component of solar variability (using monthly SSN) of SLP (in hPa) during DJF. The periods considered are 1856–1957 (**a**) and 1958–1997 (**b**). Other independent factors are OD, ENSO and trend

Now let us focus on Fig. 8.11, which is the same as Fig. 8.7 but only considers period during the latter half of twentieth century. It is seen solar cycles are relatively stronger. That period is also the overlapping period of all three studies (vL07, W97, WL08). Now Fig. 8.9 (left) suggests high solar years indicate warming, while low solar years are cooling in tropical Pacific. It is a warming signal that is captured in the EOF technique. ENSO usually follows 2–3 years cycle, and solar maximum to minimum years cover 5 years. It explains phase locking of ENSO and Sun as shown byWL08. WL08 that considers composite of 100 years record mainly captures strong ENSO features of last few solar cycles. Hence, the analyses suggest consistency among vL07, W97 and WL08 during an overlapping period, which is not true before the 1950s.

Figure. 8.12 shows results of multiple linear regression on SLP due to SSN during DJF. The right-hand side is for the period 1958–1997 and the left hand for 1856–1957. Other independent factors are the OD, trend and ENSO. It suggests solar signal on SLP also indicates differently during 1958–1997. No signal is seen around ITCZ of tropical east Pacific during the later period, and the signal around Aleutian Low is also weakened. Based on the above discussions, some mechanism relating to the Sun and ENSO may be proposed.

8.4   Proposed Mechanism: Earlier Period

MLR analysis for DJF (DJF mean value represents the year), during the earlier period, detected a solar signal in SLP, around the eastern tropical Pacific and northern Pacific (Fig. 8.12a). A significant signal though small in magnitude is detected in the tropical east Pacific that could intensify the ITCZ during higher solar (HS) years and responsible for enhancement of the trade wind. It can subsequently be accounted for uplifting the thermocline at the eastern coast of Pacific. Hence during HS years, it can produce a situation similar to that of a cold phase of the ENSO and vice versa during low solar (LS) years. A chain of a mechanism may be initiated, similar to that of the ENSO via inciting the trade wind around the ITCZ. It is originated via a solar forcing on SLP during DJF.

A wind forcing mechanism around the equatorial Pacific is required to trigger the ENSO cycle, the source being unclear. This study indicates that the Sun might influence this trigger. Such observation also suggests a decadal signal in the Walker circulation (simultaneously, with the ENSO), related to the Sun, to which some other scientists (e.g. Vecchi and Soden 2007) have pointed as 'the source still unclear'.

The empirical observation is shown during the earlier period (Fig. 8.5a) that captures similar ENSO–solar relationship, though only for HS years, when say SSN is greater than 80. A possible explanation, supported by the work of Dima et al. (2005) may shed some light on why during years of LS that observation fails to indicate the reverse.

Dima et al. (2005) identified one mode of climate, originated by solar variability, which dominates upper atmospheric levels and SLP. However, at the surface, they detected another mode, mainly dominated by atmosphere–ocean coupling that explains around three times more variance than that due to the solar. Thus, ENSO, measured in terms of the equatorial Pacific SST, which is generated via coupling in ocean–atmosphere around that place, possesses the potential to overpower the effect of solar mode.

On the other hand, during HS years, the solar signal might make its presence felt differently. It possesses the potential strength to overshadow the usual ENSO interannual behaviour. Hence, this study in the earlier period indicates that the Sun during HS years may influence equatorial Pacific SSTs, but at periods of LS activity, it is overwhelmed by the inherent strong variability of ENSO (Fig. 8.5a). Nevertheless, the decadal signal in the ENSO, as described above, originated via the solar variability, cannot be ignored. Such indication may also shed light regarding some unexplained characteristics of the ENSO (viz. prolonged lifetime or premature cessation of some ENSO cycles).

The regression fails to detect any noticeable solar signal in the eastern equatorial Pacific SST, without or with the ENSO as an independent parameter (Fig. 8.10) during an earlier period. A bias towards a cold event side of the ENSO during HS years is captured for a very few data points, in comparison to the whole set of observation (shown in Fig. 8.5a). It is the reason why regression fails to detect any robust result for SST in the tropics and the Sun. Moreover, in the regression with an inclusion of the ENSO as an index, we still fail to capture any detectable solar signal in SST. It indicates again that it is the ENSO which is most of the time dominating the eastern tropical Pacific SST. Thus, during the earlier period, the apparent influence of the Sun on ENSO may be initiated through triggering the trade wind, though may not be detected by the method of regression in equatorial SSTs.

In midlatitudes, around the AL, a strong solar signal in SLP (~6 hPa) is captured. Such a signature can be linked to a weakening of the Ferrel cell during HS years around the north Pacific and vice versa during LS years. The tropics in Pacific and midlatitude (north) are tied through ocean pathways, and we find that the solar signal in the lower troposphere is also coupled with the ocean around those places, with the Sun being a potential driving factor to modulate the coupling mechanism.

For a solar influence on the troposphere, two fundamentally different routes have been proposed: the first is generated through the stratosphere, while the other suggests the Sun directly impacts SST without the feedback from stratosphere (Meehl et al. 2008, 2009). In a GCM study (in an absence of ocean), Haigh (1996, 1999) observed a weakening of the Ferrel circulation, for an increase in solar variability at the lower stratosphere. As ocean gyres are driven by winds, such a change should be reflected in the subtropical gyre around north Pacific. Weakening of this gyre will augment temperature around that place via impeding overturning in the northeastern part of Pacific (thus mixing with cold water) (also seen in Fig. 8.10). Thus, the signature of the Sun around the northern Pacific can be transported to the tropics of Pacific and vice versa. It was mentioned earlier the Pacific is connected between the midlatitude (north) and tropics via the shallow MOC. Thus the Sun may impact not only the trade winds but also the tropics via the midlatitudes, through such a linkage.

In the other route, without any stratospheric feedback, the Sun can directly influence SST. Here we suggest that this pathway is possibly involved with the shallow MOC and originated at northern Pacific (Fig. 7.1). During HS years, the shallow MOC, by absorbing more heat around the places of northern Pacific, can weaken the AL – which in turn, can cause reduction in the strength of Ferrel circulation around that place. Through the shallow conveyor belt, the heat absorbed can again be transported to the tropical Pacific.

Thus, this study reveals that the tropics in the Pacific and midlatitude (north) are not only tied through ocean pathways, but the solar signal in the lower troposphere is also coupled with the ocean around those places. The Sun could control such a coupling mechanism and be a potential driving factor to characterise the decadal fluctuations observed in some of the parameters mainly regulated by the surface layer of the ocean. During the last 50 years, using the hydrographic historical data, a study observed decadal variability in the shallow MOC (Zhang et al. 2006). The strength of equatorward convergence across 9°N and 9°S of the pycnocline volume transport characterises a variability which is also decadal in nature. This circulation fluctuates significantly on decadal scales (Zhang et al. 2006), with decade-to-decade maximum variations about the trend is ~7–11 Sv.

8.5 Proposed Mechanism: Later Period

The solar signal on SLP is missing around the tropics, during the later period (DJF), with a significant but weakened signature at the place of AL. We now consider how the solar cycle can have a different effect on ENSO during the later period. During the earlier period (HS years), the solar signal is responsible for cooling in the tropics alongside warming in the north Pacific. Such a high-temperature gradient can hasten equatorward convergence by enhancing the rate of flow of shallow ocean current from the midlatitude (north) of Pacific towards tropics. But, that signature is weakened in the midlatitude and missing around the tropics (later period), indicates that during 1950s to 1998, the modified solar signal may have a part in reducing the strength of shallow overturning current. Such reduction is noted by several studies (Vecchi and Soden 2007; Zhang et al. 2006). During the said period, the presence of a weakened solar signature in the midlatitude may be responsible for the decadal signal still noticed in the shallow MOC. The MLR technique also suggests that unlike the earlier one (Fig. 8.10), warming in the tropical Pacific is observed around the east coast during the later period (Fig. 8.9a). It is also seen that if the ENSO signal is not excluded from the solar, then regression results for the latter half of the twentieth century suggest that the solar signal resembles that of the warm event of ENSO, which is not the case during the earlier period. Hence in the tropics, the solar signal during the later period is mixed up with ENSO (Fig. 8.9a, b). Thus during 1958–1997, an omission of ENSO from the regression gives a false signal of cooling (warming) in SST (as observed by White et al. (1997)) related to lower(higher) solar activity. We now discuss a plausible pathway for transporting such a solar signature and also the associated implications.

During the latter period, the signature of the Sun around the ITCZ is no longer present. However, during HS years, weakening of the Ferrel circulations subsequently warms the regions of northern Pacific (alongside, shallow oceanic MOC also absorbs more heat around there). It may cause warming in the tropical Pacific via the shallow ocean pathway. In Fig. 7.3, it was showed that oceanic heat content in the Pacific basin around tropics is correlated with the ENSO index – with more heat indicates a warm phase of the ENSO. Such a pathway for a signature of the Sun can transport heat from the midlatitude to tropics, which during HS years may activate a warm phase of ENSO in the tropics. A shift in the ENSO, towards a warm phase, during the latter period of HS years, is also indicated in Fig. 8.5b, in contrary to the earlier observation. Thus, in the regression, the signature of the Sun is shown to be mixed up directly with the ENSO, during the latter period, producing a comparable influence on SST in tropics (Figs. 8.9a, b).

Thus, the Sun at HS years, during the earlier period, could be a potential factor for triggering a cold ENSO phase, may produce a contrary influence during the latter period. Compared to the earlier period, such a bias of HS years on the ENSO warm side during the latter was also identified in the scatter plot of Fig. 8.5b, compared to 8.5a, though max solar years (used by vL07) are still observed to be on the ENSO cold phase. Relating to solar–ENSO behaviour, apart from indicating a plausible mechanism, these explanations/results are also able to reconcile few previous contradictory findings.

8.6   Contradictions and Reconciliations

The regression technique of the 1958–1997 period can explain some of the inconsistencies relating to the signature of the Sun around the east tropical Pacific, for instance, the inconsistent findings of White et al. (1997) and vL07, in addition to White and Liu (2008) as shown in Fig. 8.6. During the period, common to all these works, the regression study, accompanied by the observation that the peak SSN years used by vL07 falls before the maximum of the smoothed DSO by a year or more (Fig. 8.11), gives coherence to these seemingly conflicting results.

First, it was identified that to detect the solar signal by vL07, the methodology adopted is not characterising the actual behaviour of a solar QDO. In fact, peak years, as used by vL07, reflect a different (rising) phase of the solar cycle to peak year of smoothed DSO because it falls a year or more in advance of the broader maximum of the solar 11-year cycle (Fig. 8.7). It was noted that the solar SSN peak years are usually allied with the ENSO cold events (Fig. 8.2). As it takes usually 1–2 years for the ENSO to alter between phase (cold to warm), this explains why the observations of White et al. (1997) differ to that from vL07. Omission of ENSO, during 1958–1997, from the regression gives a false cooling (warming) signature on SST in the tropics for lower (higher) solar activity (Fig. 8.9). It clearly indicates why during the period, common to all the studies, both White and Liu (2008) and White et al. (1997) not only detect warming in HS years but also cooling in LS years (Fig. 8.6). Throughout the solar cycle, the ENSO peak and trough show a period of about 2 years (as detected by White and Liu (2008)) due to its usual phase transition mechanism(Fig. 8.6, bottom). Hence, during the common period to all studies, there is a consistency between White and Liu (2008), White et al. (1997) and vL07 (Fig. 8.6). It also explains, during 1958–1997, why max solar years (as used by vL07) are still on the cold phase of ENSO (shown in Fig. 8.5, bottom), which is simply an artefact of the ENSO and not any particular features of the rising phase of solar cycles.

Moreover, this compositing study during 1900–2000, as White and Liu (2008) considered, generally captures the behaviour for period 1958–1997, the time strongly affected by the ENSO and having stronger solar cycles. It would be unable to detect similar phase locking of the solar and ENSO, if the application of their approach was considered to the time before 1958. Additionally, it shows that during the later period mixing of the ENSO signal with the Sun might modulate the actual signature of the Sun (and hence insensitive to some methodology) and needs to be accounted carefully.

References

Chen D, Cane MA, Kaplan A, Zebiak SE, Huang D (2004) Predictability of El Niño over the past 148 years. Nature 428(6984):733–736

Dima M, Lohmann G, Dima I (2005) Solar-induced and internal climate variability at decadal time Scales. Int J Climatol 25:713–733. https://doi.org/10.1002/joc.1156

Haigh JD (1996) The impact of solar variability on climate. Science 272(5264):981–984

Haigh JD (1999) A GCM study of climate change in response to the 11-year solar cycle. Q J Roy Meteorol Soc 125(555):871–892

Hamm E, Tung KK (2012) Statistics of solar cycle–La nina connection: correlation of two autocorrelated time series. J Atmos Sci 69(10):2934–2939

Jacob R, Schafer C, Foster I, Tobis M, Anderson J (2001, April) Computational design and performance of the fast ocean atmosphere model, Version One, In: Alexandrov VN, Dongarra JJ, Tan CJK (eds) Proceedings 2001 international conference on computational science, Springer, pp 175–184, Also ANL/CGC-005-0401

Kodera K, Coughlin K, Arakawa O (2007) Possible modulation of the connection between the Pacific and Indian Ocean variability by the solar cycle. Geophys Res Lett 34(3)

Krivova NA, Solanki SK, Fligge M, Unruh YC (2003) Reconstruction of solar irradiance variations in cycleÂ 23: is solar surface magnetism the cause? Astron Astrophys 399(1):L1–L4

Lean J (2000) Evolution of the Sun's spectral irradiance since the maunder minimum. Geophys Res Lett 27(16):2425–2428

McPhaden MJ, Zhang D (2004) Pacific ocean circulation rebounds. Geophys Res Lett 31:L18301. https://doi.org/10.1029/2004GL020727

Meehl GA, Arblaster JM, Branstator G, van Loon H (2008) A coupled air-sea response mechanism to solar forcing in the pacific region. J Climate 21(12):2883–2897

Meehl GA, Arblaster JM, Matthes K, Sassi F, van Loon H (2009) Amplifying the pacific climate system response to a Small 11-Year solar cycle forcing. Science 325:1114–1118. https://doi.org/10.1126/science.117287

Roy I (2010) Solar signals in sea level pressure and sea surface temperature. Department of Space and Atmospheric Science, PhD Thesis, Imperial College, London

Roy I (2014) The role of the sun in atmosphere-ocean coupling. Int J Climatol https://doi.org/10.1002/joc.3713

Roy I, Haigh JD (2010) Solar cycle signals in sea level pressure and sea surface temperature. Atmos Chem Phys 10(6):3147–3153

Roy I, Haigh JD (2012) Solar cycle signals in the pacific and the issue of timings. J Atmos Sci 69(4):1446–1451. https://doi.org/10.1175/JAS-D-11-0277.1

Roy I, Asikainen T, Maliniemi V, Mursula K (2016) Comparing the influence of sunspot activity and geomagnetic activity on winter surface climate. J Atmos Sol Terr Phys 149:167–179

van Loon H, Meehl GA, Shea DJ (2007) Coupled air-sea response to solar forcing in the Pacific region during northern winter. J Geophys Res-Atmos 112:D02108. https://doi.org/10.1029/2006JD007378

Vecchi GA, Soden BJ (2007) Global warming and the weakening of the tropical circulation. J Climate 20:4316–4340

White WB, Liu ZY (2008) Non-linear alignment of El Nino to the 11-yr solar cycle. Geophys Res Lett 35:L19607. https://doi.org/10.1029/2008GL034831

White WB, Lean J, Cayan DR et al (1997) Response of global upper ocean temperature to changing solar irradiance. J Geophys Res-Oceans 102(C2):3255–3266

Zhang D, McPhaden MJ (2006) Decadal variability of the shallow Pacific meridional overturning circulation: relation to tropical sea surface temperatures in observations and climate change models. Ocean Model 15(3-4):250–273

Zhang Y, Wallace JM, Battisti DS (1997) ENSO-like interdecadal variability: 1900–93. J Clim 10:1004–1020

Chapter 9
A Debate: The Sun and the QBO

Abstract This chapter initially discussed results of data analysis on solar and QBO separately. It is followed by a discussion on two very popular research relating to the combined influence of the Sun and QBO on the upper polar stratosphere. It reconciled why these two studies differ with each other at a certain point. Later the combined effect of the Sun and QBO was investigated on surface climate. The knowledge can be used for improving prediction skill in the polar region.

Keywords QBO · Stratospheric Polar temperature · Polar Vortex · Solar Flux 10.7cm · Brewer–Dobson circulation · Solar*QBO index

Labitzke and van Loon (1992) plotted a scatter diagram (shown in Fig. 3.4 and also discussed in Sect. 3.5) that depicts values of solar 10.7 cm flux along the abscissa and 30 hPa temperature at the north pole for each year in Feb/Jan as ordinate. The QBO phase was drawn with symbols, square for westerlies and triangle for easterlies. The demarcating vertical and horizontal line have been marked to indicate regions where certain phases of the QBO predominate and shown by the E and W labels. From that figure, it is clear that warm polar temperatures tend to occur during the east phase of QBO at solar minimum and west phase at solar maximum. Whereas the cold polar temperature occurs during W-ly QBO at solar minimum and E-ly QBO at solar maximum.

Why polar temperature in the upper stratosphere is important? As Kodera and Kuroda (2002) discussed, it indicates the strength of Brewer–Dobson circulation (as shown in Fig. 3.7 and discussed in details in Sect. 3.8) can affect lower stratosphere. Finally, perturbation in lower stratosphere has implication in the circulation fields of the troposphere as examined in various studies by Haigh (2003) and Haigh et al. (2005) and also shown in Fig. 9.1.

Haigh et al. (2005) using a simplified global circulation model (GCM) showed that despite the lack of a stratospheric polar vortex and with the presence of only a uniform stratosphere, it is possible to reproduce patterns of the troposphere. This model did not have oceans. It is done by using perturbations in broad latitudinal scale. Such a study indicates that for understanding the tropospheric effects of solar influence, a detailed representation of the stratosphere is not essential, although the

© Springer International Publishing AG, part of Springer Nature 2018
I. Roy, *Climate Variability and Sunspot Activity*, Springer Atmospheric
Sciences, https://doi.org/10.1007/978-3-319-77107-6_9

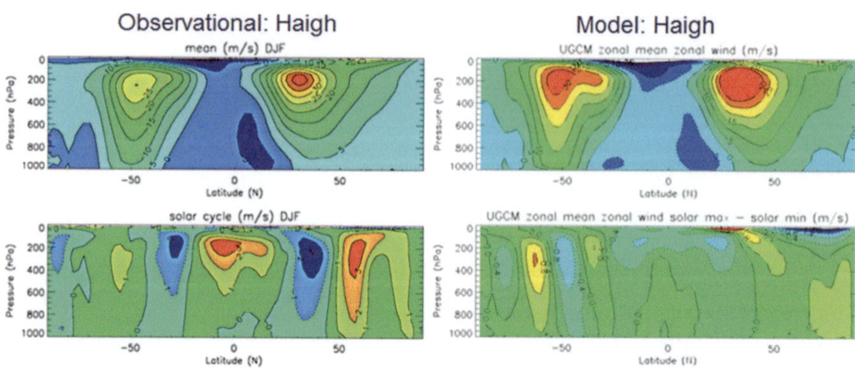

Fig. 9.1 Zonal mean zonal wind (u) plot as a function of pressure and latitude (top); the difference between maximum and minimum solar phase for u (below). The right hand is a model result and left hand from observation using NCEP Reanalysis data. (After Haigh and Blackburn 2006)

heating source in the stratosphere remains a critical factor. Haigh and Blackburn (2006) in their simplified atmospheric general circulation model shows that imposed changes of temperature forcing in the lower stratosphere lead to coherent changes in the midlatitude jet stream and its associated storm track in terms of latitudinal width and location. They showed that feedbacks from eddy/mean flow are crucial to such variations. Hence, solar heating of the lower stratosphere may alter the tropospheric circulation without even any direct forcing from underneath the tropopause. Their observational analysis using multiple regression of NCEP Reanalysis data also captured such tropospheric changes (Fig. 9.1, left).

Polar temperature around stratopause can also potentially affect polar lower atmosphere through downwards propagation of polar annular modes as observed by Baldwin and Dunkerton (2001) (as shown in Fig. 3.5 and discussed in details in Sects. 3.6 and 3.7).

9.1 Data Analysis: Solar and QBO Separately

Haigh (2003) using the multiple regression technique detected independent signals of the Sun and QBO in the zonal mean temperature. Their result capturing amplitudes of the component of zonal mean temperature variability due to QBO (at the top) and solar (bottom) is shown in Fig. 9.2. Here significant regions up to 95% are marked by white. From Fig. 9.2, it is evident that no significant signal for the solar or QBO is seen in the north polar temperature around 30 hPa level (~30 km) if considered separately. However, data sorting of the solar and QBO indicates differently. It suggests that it is important to segregate the meteorological data based on the QBO phase, to detect a clear 11-year solar signal (Labitzke and van Loon (1992) in polar stratosphere, also shown in Fig. 3.4 and discussed in Sect. 3.5).

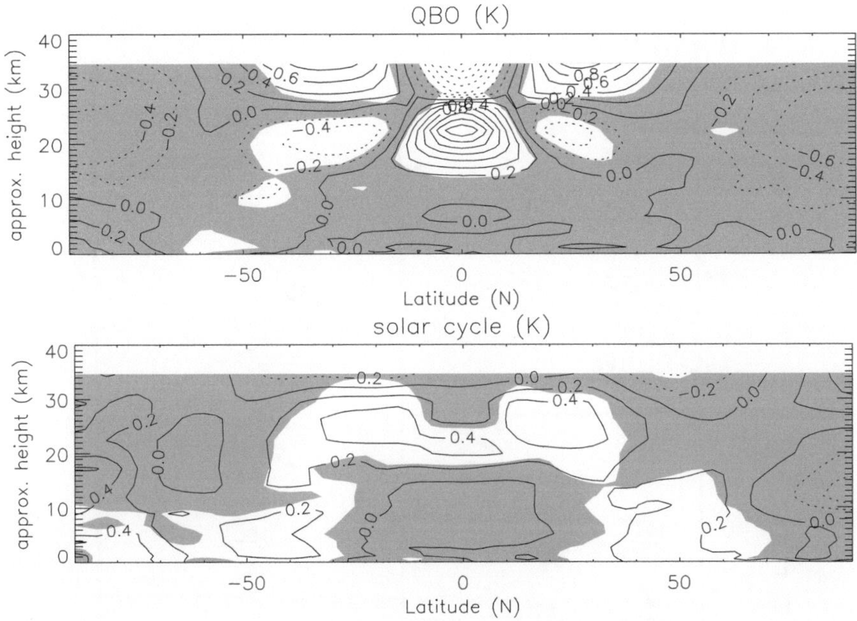

Fig. 9.2 Amplitudes of the component of zonal mean temperature variability for: (above) QBO and (below) solar. Significant regions up to 95% are marked by white. (After Haigh 2003)

A recent analysis of Camp and Tung (2007) using slight different temporal and spatial coverage of winter north polar temperature, however, indicates certain contradiction to that from Labitzke and van Loon (1992) relating to solar/QBO relationship. They applied linear discriminant analysis on NCEP–NCAR reanalysis data for 51 years of north polar temperature during late winter (Feb–Mar). Regarding north polar temperature, they considered the 10–50 hPa layer mean temperature, as calculated by the difference between 50 and 10 hPa geopotential height surfaces. The summary of their findings is shown in Fig. 9.3b in a four-quadrant diagram, with solar flux 10.7 cm as abscissa and QBO (m/s) as ordinate. Following the usual convention, westerly QBO is shown with positive values whereas easterly with negative. The demarcating horizontal and vertical lines have been drawn to indicate phases of the QBO in relation to solar max or min, thus comprising a total of four groups, viz. Solar-max/E-QBO, Solar-min/E-QBO, Solar-max/W-QBO and Solar-min/W-QBO. The arrows indicate the direction of warming. According to the Fig. 9.3b, it is clear that though the state of the QBO westerly phase during solar minimum emerges as a distinct coldest state like Labitzke and van Loon (1992) (as shown in Fig. 9.3a), but the easterly phase of QBO during solar maximum emerges very warm indeed. Here lies the contradiction, and hence, our initial attempt is to understand the main reason of such inconsistency between the two findings (Labitzke and van Loon 1992; Camp and Tung 2007), involving the polar temperature, QBO and Sun.

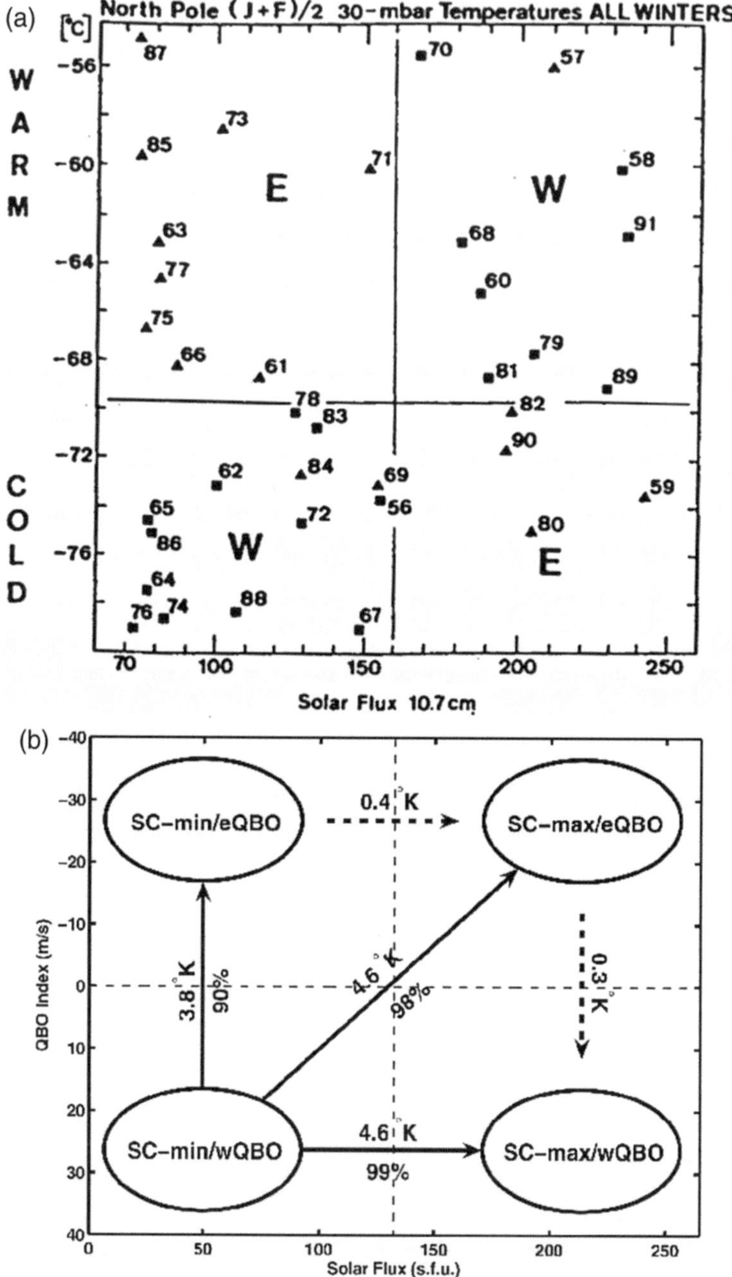

Fig. 9.3 Solar–QBO relationship: (**a**) Labitzke and van Loon (1992) and (**b**) Camp and Tung (2007)

9.2   Polar Temperature During JF with Respect to QBO (40 hPa) and F10.7

We started this work keeping as close as possible to the two work as mentioned above using north polar temperature from NCEP–NCAR (National Centres for Environmental Prediction–National Centre for Atmospheric Research) reanalysis product, between the period 1953–2001. First, we considered similar temporal (JF) and spatial coverage (30 hPa) of north polar temperature dataset as used by Labitzke and van Loon (1992). Following them, we used QBO at 40 hPa level (in m/s) and solar variability with solar 10.7 cm flux. A scatter plot (Fig. 9.4a) was produced with QBO (40 hPa) as ordinate and solar 10.7 cm flux along the abscissa. Since Salby and Callaghan (2004) observed that there is a change in sign of QBO winds during the winter of some max solar years, Camp and Tung (2007) used an average value of the QBO during DJFM (Dec-Jan-Feb-Mar) to reduce monthly fluctuation. Here, in this analysis, we used the mean value of DJFM (Dec-Jan-Feb-Mar) for both the solar F10.7 and QBO. A demarcating horizontal line has been drawn to separate the E-ly (shown with a negative value) to that from the W-ly (shown with positive value). Whereas a demarcating vertical line has been drawn above 155 of solar flux F 10.7 (like Labitzke and van Loon 1992), to segregate higher solar years to that from the lower solar years. These vertical and horizontal line generate a total of four groups representing a various combination of the solar and QBO, viz. Solar-min/W-QBO, Solar-min/E-QBO, Solar-max/E-QBO and Solar-max/W-QBO. Polar temperatures anomalies with sufficient high value (say, greater than 11.5 K (chosen, to indicate good contrast in schematic)) have been marked by red colour. From Fig. 9.4a, it is seen that the quadrant where almost all the points are red lies in the box representing solar max/W-QBO, whereas the quadrant with least population of red spots represents solar min/W-QBO – thus consistent with Labitzke and van Loon (1992).

To reproduce the same work in a form Fig. 9.3b like Camp and Tung (2007), we presented the work as mentioned earlier in a four-quadrant box (Fig. 9.4b), where each quadrant represents the same state of solar and QBO relationship as demarcated by dotted lines in Fig. 9.4a. In each box, inside an oval, we mentioned the state of the solar and QBO and showed the average value (in red) of all the polar temperatures (in K) within the respective quadrant. In calculating the mean value, years those are adversely affected by the volcano (thus excluding 1992, 1983 and 1964, as we restrict to JF) are omitted; such exclusion follows Labitzke and van Loon (1992). Arrows point from the cold to warm (likewise Camp and Tung 2007) and thus indicate the direction of warming. From Fig. 9.4b it is quite clear that the box representing solar Min/W-ly QBO shows the coldest polar temperature (with average 200.2 K). Whereas the quadrant representing solar Max/W-ly is seen to be the warmest with an average value of 208.7 K, followed by solar Min/E-ly (with mean value 207.7 K). Notably, a warming of around 8° in these two quarters in respect to solar Min/W-ly QBO is clearly noticed in Fig. 9.4b. Hence, to summarise the observations of Fig. 9.4a and b, it can be stated that using the QBO height at 40 hPa, we

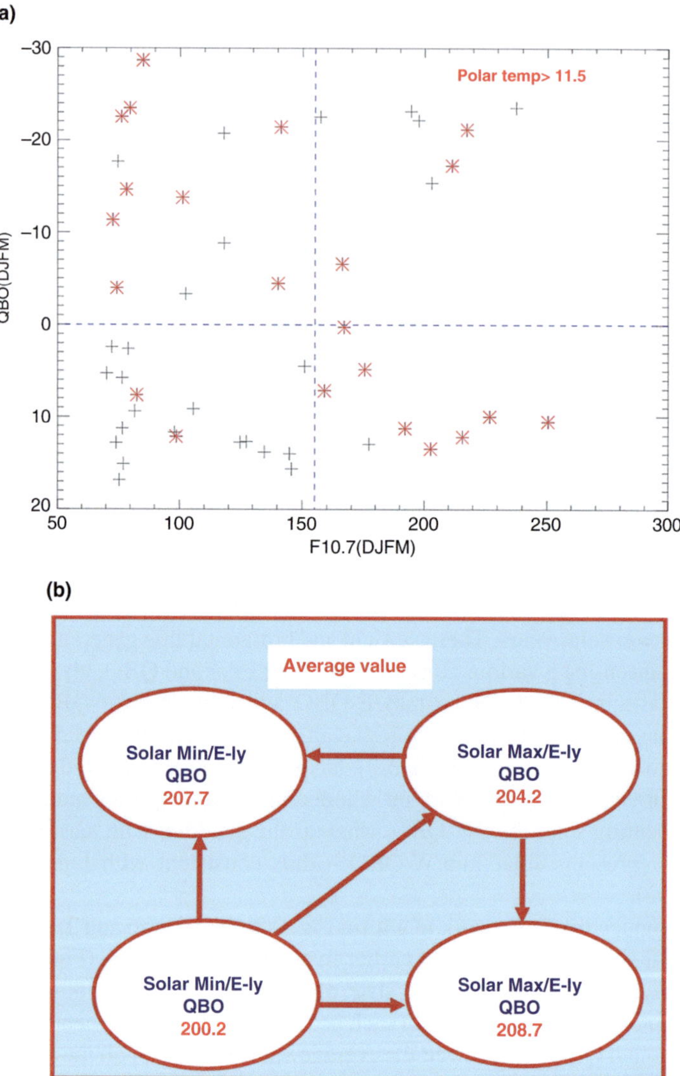

Fig. 9.4 Polar temperature during JF in different quarters of solar and QBO (40 hPa): (**a**) absolute value and (**b**) average value (Roy 2010)

notice: cold polar temperature tend to occur during W-ly QBO at solar minimum and E-ly QBO at maximum solar phase. Whereas warm polar temperatures occur during east phase of QBO at solar minimum and the west phase at solar maximum. Thus, this result is in agreement with the work of Labitzke and van Loon (1992) concerning the north polar temperature, QBO (40 hPa) and solar. In an updated version, Labitzke (2004) using QBO height 45 hPa also verified such observation.

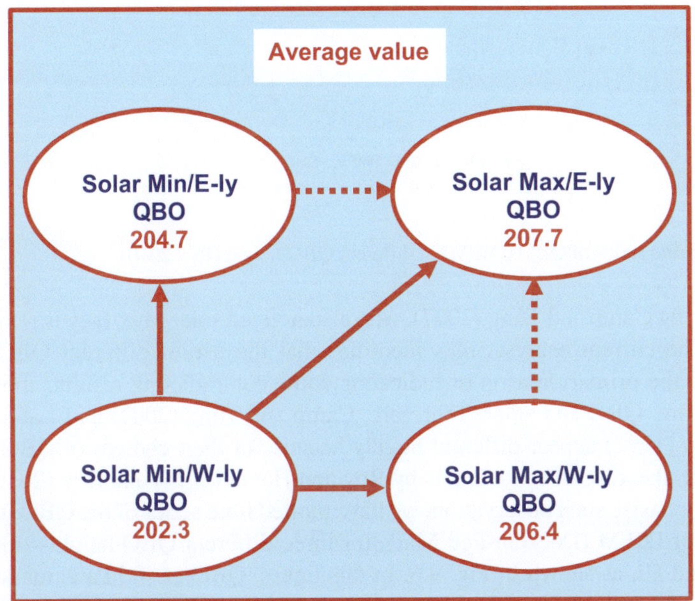

Fig. 9.5 Polar temperature (average value) during JF in different quarters of solar and QBO (30 hPa)

9.3   Polar Temperature During JF for QBO (30 hPa) and F10.7

Here, the same work is reproduced, with only one exception – in place of QBO at 40 hPa level (as used by Labitzke and van Loon 1992), the same analysis is repeated using QBO at 30 hPa level (employed by Camp and Tung 2007).

After carrying on with a similar analysis, like Fig. 9.4a (not shown here), we calculated the average value of north polar temperature during different phases of solar and QBO (**30 hPa**) combination. Such representation is depicted in Fig. 9.5 which is done in a similar manner like Fig. 9.4b. From Fig. 9.5, it is evident that the quadrant representing solar Min/W-ly is again the coldest (with average temperature 202.3 K). But, surprisingly **here Solar Max/E-ly appears to be the warmest** (with average 207.7 K), followed by solar Max/W-ly (average 206.4 K). Now, let us discuss the direction of warming, as indicated by red arrows. In Fig. 9.5, if we focus on the quarter solar Min/W-ly, we observe that arrows showing the direction of warming remain the same, as is seen in Fig. 9.4b – suggesting that this particular quarter is still the coldest of all. However, if we focus on the quarter solar max/E-ly, we surprisingly observe that the direction of two arrows around this quarter is different to that from Fig. 9.4b (hence marked by red dotted lines). Thus, only using QBO at 30 hPa level, in place of 40 hPa, is seen to have even changed the direction of warming in some quarters, suggesting solar Max/E-ly is the warmest one. Such observation

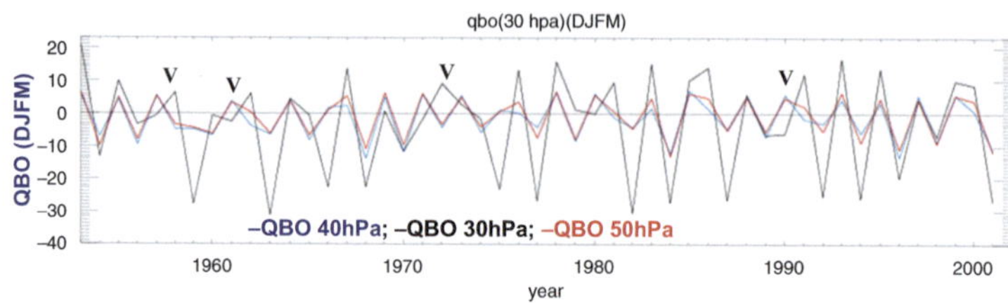

Fig. 9.6 Time series of QBO (DJFM) in different QBO height (Roy 2010)

is following Camp and Tung (2007), who also noticed solar Max/E-ly is remarkably warm. The current analysis thus identifies that the use of different QBO height might be the primary reason of indicating some contradicting results, about polar temperature, QBO and solar. Therefore, Camp and Tung (2007) and Labitzke and van Loon (1992) appear different largely because of their choices of QBO height. Those are also discussed in details by Roy and Haigh (2011) and Roy (2010).

To emphasise such observation, we have plotted time series of the QBO, using an average of DJFM (Dec-Jan-Feb-Mar), for three different QBO heights (hPa), viz. 30, 40 and 50, as shown in Fig. 9.6. In this figure, QBO at 40 hPa is marked with blue, 30 hPa with black and 50 hPa with red. From this figure, it is clear that there are some places (marked by 'V'), where the QBO at 30 hPa is seen to be out of phase with QBO at 40 hPa. That means, on occasions, QBO at 30 hPa is seen to be easterly (westerly) though westerly (easterly) using height 40 or 50 hPa level. However, interestingly, QBO at 40 or 50 hPa does not suffer much phase change, indicating that results might not be very sensitive using QBO at 40 or 50 hPa level. Thus, Labitzke (2004) using QBO height 45 hPa arrives at same result of Labitzke and van Loon (1992), who used QBO at 40 hPa.

9.4 Time Series of QBO at Different Height and EOF Analysis

As mentioned in Sect. 2.2.1, the QBO travels downward in a speed of ~1 km per month. Figure 9.7 shows time series of the QBO, at different heights (in hPa), i.e. 15, 20, 25, 30, 35, 40, 45, 50, 60 and 70. During the course of propagation apart from suffering a change of phase, they also show considerable variability, regarding amplitude. It clearly suggests that not only some results, using data sorting, are sensitive to the QBO height (as shown earlier), but also results using other techniques might be sensitive to its height. For instance, the use of different QBO heights in the regression is liable to produce different results and consequently carries different spurious implications. Thus to have an understanding of the temporal and spatial variability of the QBO, EOF technique is applied. Figure 9.8a shows the first few PC time series, whereas Fig. 9.8b depicts the first two spatial structures. PC0, which is the first principle component time series, explains 58% variance; whereas PC1, the second principal component time series, explains ~37%.

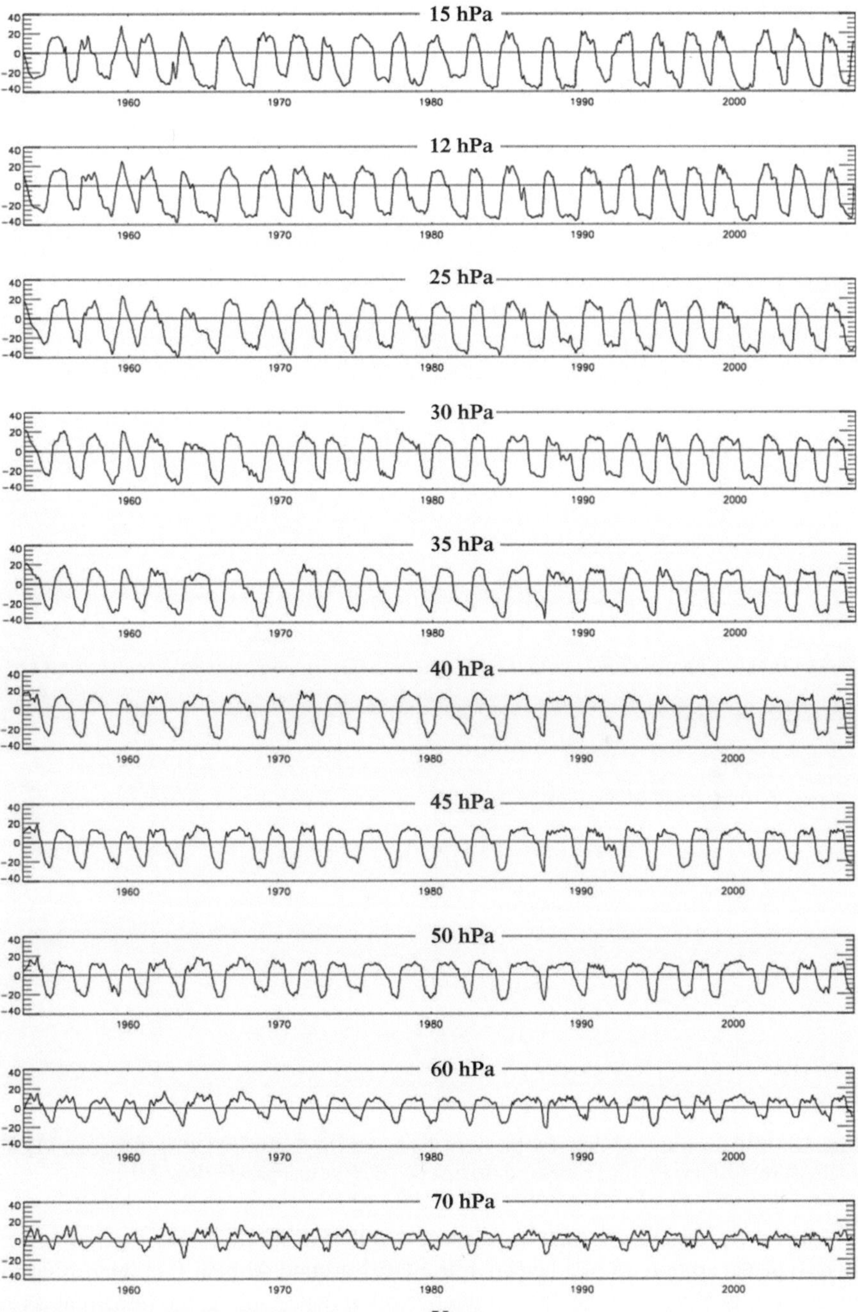

Fig. 9.7 QBO time series at different levels (hPa): 15, 20, 25, 30, 35, 40, 45, 50, 60 and 70 (Roy 2010)

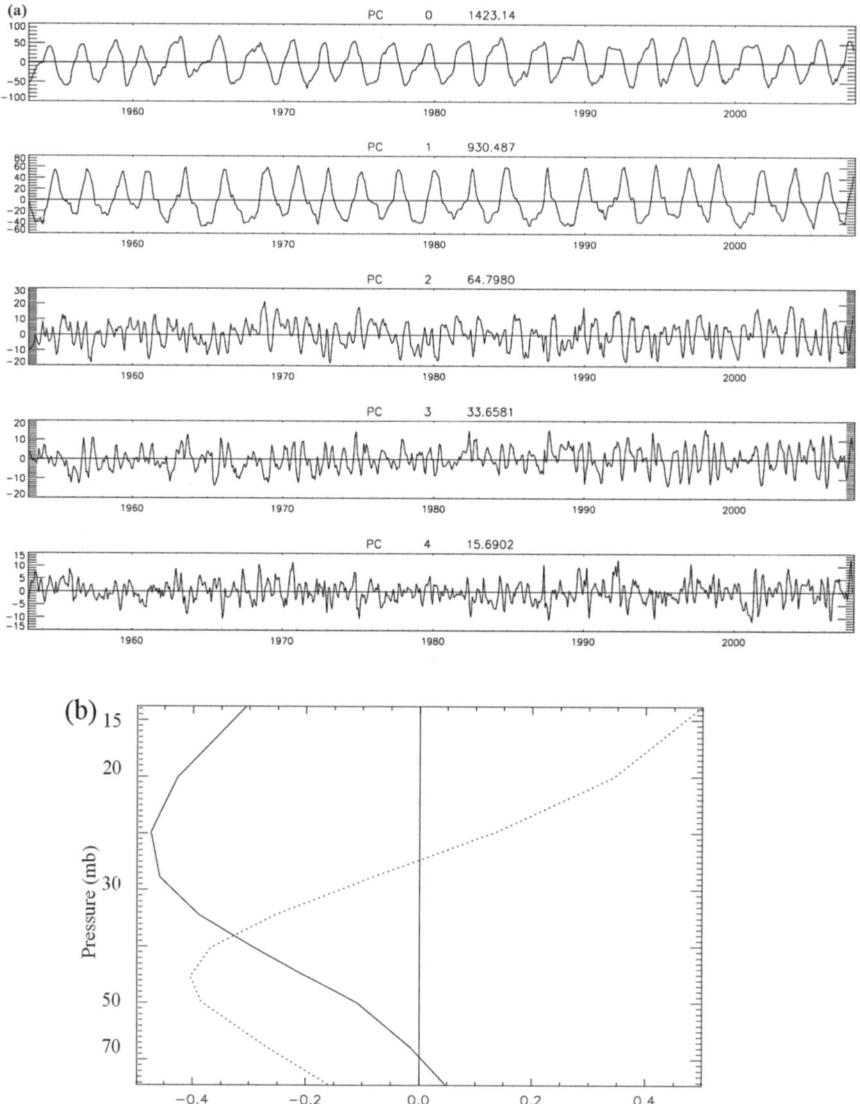

Fig. 9.8 EOF analysis: (**a**) first few PC time series and (**b**) first two EOF spatial pattern EOF analysis: (**a**) first few PC time series and (**b**) first two EOF spatial pattern (Roy 2010)

Moreover, from Fig. 9.8b, it is observed that their respective spatial patterns, i.e. EOF 0 peaks around 25 hPa level, while EOF 1 around 48 hPa. It is interesting to note: these two PCs include 95% of the overall variance so that the remaining PCs carry nominal implications and may be ignored. Moreover, being orthogonal in nature and converging approximately to the time series near 25 and 48 hPa indicate that using the time series at ~30 and 40/50hpa is equivalent to using PC0 and PC1.

9.5   Combined Effects: Solar with QBO

Using NCEP data during the second half of the last century, Haigh and Roscoe (2006), through multiple regression analysis, showed that there is no statistically significant signal of the Sun in either the surface SAM or NAM. However, when a new index that is the product of the QBO indices and solar is constructed, a good correlation is observed at the lower levels in the winter NAM and throughout the atmosphere for the SAM. Here following Haigh and Roscoe (2006), we used the index, the product of the QBO and solar in the regression which is described below.

Regression of SLP with 'Solar*QBO'
Using observational analysis, Baldwin and Dunkerton (2001) showed perturbations in the polar vortex appear to be associated with the downward propagation through polar annular modes. Such downward propagation was also identified by Thompson and Wallace (2000) in both the polar modes. In the strength of the stratospheric polar vortex, the large-amplitude variations are typically followed with a lag of less than 1 month in the tropospheric circulation by similarly signed anomalies. It persists for up to 3 months (in SH) to 2 months (in NH) (shown in Fig. 3.5 and discussed in Sect. 3.6). Based on their results, we carry the regression analysis, using all months of the year. Here we used a new index, the product of the QBO and solar (shown in Fig. 9.9) and termed as 'solar*QBO'. It is also employed in the regression analysis of Haigh and Roscoe (2006) as it well captures the behaviour of the polar temperature in terms of both the solar and QBO (following the analysis of Labitzke van Loon (1992) and Labitzke (2004)). The updated version of Labitzke (2004) even showed that the observed behaviour regarding the polar temperature, QBO and Sun is not only present in the NH but also in the SH.

Figure 9.10 shows the amplitude of the component of variability on SLP for this new index the 'solar*QBO', using other independent factors as OD, a trend and ENSO. In the analysis, QBO height 50 hPa is used. Here negative NAM and SAM features are quite evident suggesting that HS years during westerly phase of the QBO and LS years during easterly phase of the QBO, both trigger negative surface NAM and SAM features in a zonal sense. Whereas the reverse is true during HS years with the QBO E-ly phase and LS years during the QBO W-ly phase. Interestingly using 30 hPa, 40 hPa and 50 hPa, similar pattern is noticed. Such observations indeed suggest that the QBO effect (irrespective of its height at 50, 40 or 30 hPa level), in combination with the Sun reveals a negative signature in both the de-seasonalised polar annular modes. It is in agreement with Haigh and Roscoe (2006), who in their index 'solar*QBO', using QBO at 40 hPa level, also arrive at a similar sense of zonal features regarding polar annular modes. In the regression, in place of the 'solar*QBO' time series, we have also used solar*QBO(PC0) and solar*QBO(PC1) time series (not shown here). As expected, the former gives similar results using solar*QBO(30), while the latter using solar*QBO(40). A recent study (Roy et al. 2016) also investigated solar*QBO signal on SLP using a different proxy for solar indices. It again found a strong influence around polar regions during northern winter.

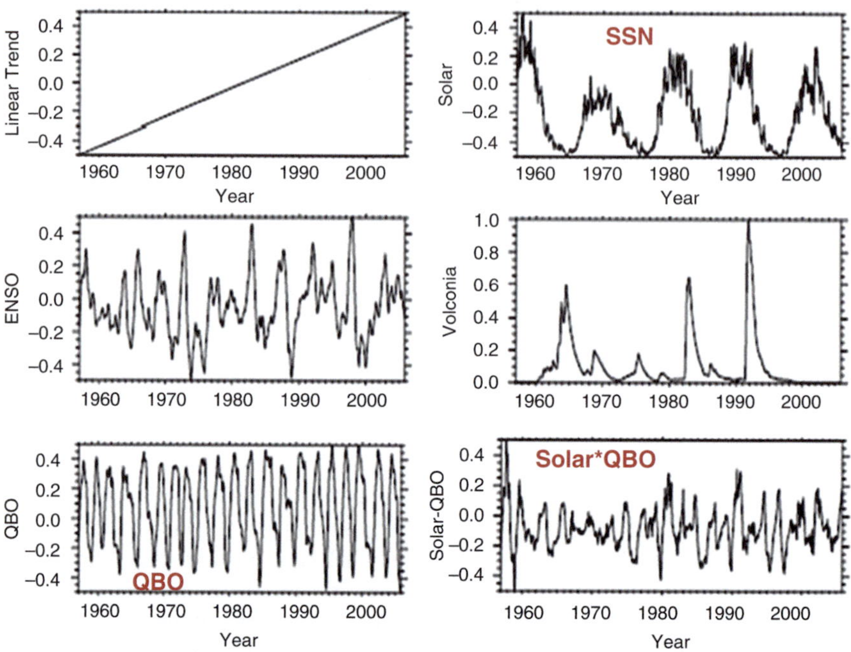

Fig. 9.9 Time series of Trend, ENSO, QBO, SSN and Volcano. A new index solar*QBO is formed as shown and used instead of SSN and QBO

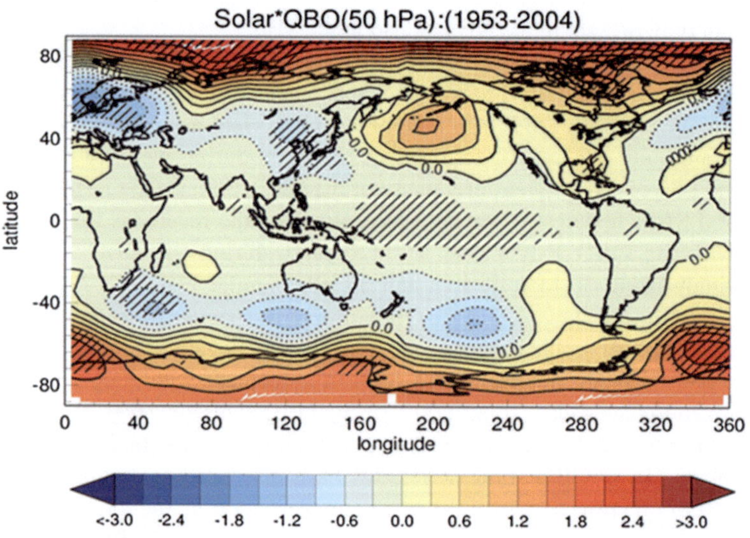

Fig. 9.10 Results of multiple regression technique (1953–2004) of SLP. Components (hPa) due to the solar*QBO (at 50 hPa) compound signature using other independent factors as OD, trend and ENSO (Roy 2014)

The recent publication of Brönnimann et al. (2007), introducing a QBO reconstruction that dated back to 1900, enables to examine its role over a longer term climate data. Repeating same analysis and combining the QBO and solar indices show that the general behaviour observed during 1953–2004 also prevails during the first 50-year period (1900–1952) in the SH, irrespective of the QBO height.

9.6   Summary

The first discussion was about middle atmosphere coupling describing the role of both the QBO and the Sun in relation to polar stratospheric temperature. Following different studies of Baldwin and co-workers (2001, 2005) and Thompson and Wallace (2000) those indicate perturbations in the polar stratosphere propagate downwards, affecting even the lower troposphere over the next few months, we subsequently focused to detect the combined effect of QBO and solar, in the lower troposphere.

However, the first attempt was to understand the main cause for the apparent discrepancy relating to the polar temperature, the Sun and the QBO between two published research. Using a slightly different coverage of spatial and temporal winter polar temperature data, Labitzke and van Loon (1992) and Camp and Tung (2007) arrives at some contradictory findings. According to Labitzke and van Loon (1992), solar max/QBO E-ly is cold, while Camp and Tung (2007) detect the same as very warm. Our analysis indicates that they appear different largely because of using a different QBO height. The QBO at 30 hPa (used by Camp and Tung 2007) is seen to be out of phase with QBO 40 hPa (employed by Labitzke and van Loon 1992) in some times. That means, on occasions, QBO at 30 hPa is seen to be easterly though westerly using height 40 or 50 hPa level and vice versa. Hence, the use of different QBO height appears to be the most likely cause to influence the results of data analysis. However, interestingly, between 40 and 50 hPa, the QBO does not suffer much phase change, indicating that the results might not be very sensitive using QBO at either 40 or 50 hPa level. Since results appeared susceptible to QBO height, to understand the major variability patterns of the QBO, temporally as well as spatially, we carried out an EOF analysis. Our analysis suggests, PC0, the first principal component time series of QBO explains ~58% of the overall variance; whereas the PC1 (the second component) explains ~ 37%. Moreover, their respective spatial patterns, viz. EOF 0 and EOF1 peaks around 25 hPa and ~48 hPa level, respectively. Being orthogonal in nature it indicates that using the time series at ~30 and 40/50hpa is equivalent to using PC0 and PC1.

A different analysis of Baldwin and co-workers (2001, 2005) showed that fluctuations in the polar vortices strength of stratosphere are coupled downward to surface climate in both hemispheres. The large-amplitude variations in the vortex strength are typically followed by similarly signed anomalies in the tropospheric circulation with a lag of less than 1 month that persists for up to 2 months (in NH) to 3 months (in SH). To address this, we carried out the MLR analysis on SLP, where the regression was done using all months of the year, using other independent factors as OD and trend.

Following Baldwin and Dunkerton (2005) who speculated that the pathway involving the polar modes of variability seems to consider interactions of the solar with QBO, we used a new index in our multiple regression analysis. Here in place of the solar index, a new index is used – the product of QBO and solar (and termed as 'solar*QBO'), to represent the collective behaviour involving both the solar and QBO as observed by Labitzke and van Loon (1992) and Labitzke (2004). Such index has also been used by Haigh and Roscoe (2006) in their multiple regression analysis. It suggests that irrespective of the QBO height of 30, 40 or 50 hPa level its effect in combination with the Sun reveals a negative signature in the de-seasonalised NAM and SAM for the last 50 years period. That indicates that HS years during the QBO W-ly phase and LS years during the QBO E-ly phase, both trigger negative NAM and SAM features, in the zonal mean sense; whereas the reverse is true during HS years with the easterly QBO phase and LS years during the westerly QBO phase.

References

Baldwin MP, Dunkerton TJ (2001) Stratospheric harbingers of anomalous weather regimes. Science 294(5542):581–584. https://doi.org/10.1126/science.1063315

Baldwin MP, Dunkerton TJ (2005) The solar cycle and stratosphere-troposphere dynamical coupling. J Atmos and Sol-Terr Phys 67:71–82. https://doi.org/10.1016/j.jastp.2004.07.018

Brönnimann S, Annis JL, Vogler C, Jones PD (2007) Reconstructing the quasi-biennial oscillation back to the early 1900s. Geophys Res Lett 34:L22805. https://doi.org/10.1029/2007GL031354

Camp CD, Tung KK (2007) The influence of the solar cycle and QBO on the late-winter stratospheric polar vortex. J Atmos Sci 64(4):1268–1283. https://doi.org/10.1175/JAS3883.1

Haigh JD (2003) The effects of solar variability on the Earth's climate. Philos T R Soc A 361(1802):95–111

Haigh JD, Blackburn M (2006) Solar influences on dynamical coupling between the stratosphere and troposphere. Space Sci Rev 125:331–344

Haigh JD, Roscoe HK (2006) Solar influences on polar modes of variability. Meteorol Z **15**(3):371–378. https://doi.org/10.1127/0941-2948/2006/0123

Haigh JD, Blackburn M, Day R (2005) The response of tropospheric circulation to perturbations in lower-stratospheric temperature. J Climate 18(17):3672–3685

Kodera K, Kuroda Y (2002) Dynamical response to the solar cycle. J Geophys Res 107(D24):4749. https://doi.org/10.1029/2002JD002224

Labitzke K (2004) On the signal of the 11-year sunspot cycle in the stratosphere and its modulation by the quasi-biennial oscillation. J Atmos and Sol-Terr Phy 66:1151–1157. https://doi.org/10.1016/j.jastp.2004.05.011

Labitzke K, van Loon H (1992) On the association between the QBO and the extratropical stratosphere. J Atmos Terr Phys 54(11/12):1453–1463

Roy I (2010) Solar signals in sea level pressure and sea surface temperature. Department of Space and Atmospheric Science, PhD Thesis, Imperial College, London

Roy I (2014) The role of the Sun in atmosphere-ocean coupling. Int J Climatol 34 (3):655–677

Roy I, Haigh JD (2011) The influence of solar variability and the quasi-biennial oscillation on lower atmospheric temperatures and sea level pressure. Atmos Chem Phys 11(22):11679–11687

Roy I, Asikainen T, Maliniemi V, Mursula K (2016) Comparing the influence of sunspot activity and geomagnetic activity on winter surface climate. J Atmos Sol Terr Phys 149:167–179

Salby M, Callaghan P (2004) Evidence of the solar cycle in the general circulation of the stratosphere. J Climate 17(1):34–46

Thompson DWJ, Wallace JM (2000) Annular modes in the extratropical circulation. Part I: month-to-month variability. J Climate 13:1000–1016

Chapter 10
Solar Influence: 'Top Down' vs. 'Bottom Up'

Abstract Two fundamentally different routes have been proposed for a solar influence on the troposphere: the first is the 'bottom-up' and the other the 'top-down' mechanism. In the 'bottom-up' pathway, the Sun can directly influence sea surface temperature (SST) without stratospheric feedback, whereas the 'top-down' solar influence is generated through the stratosphere without any influence from oceans. Those two routes are discussed in details with relevant supporting studies.

Keywords Solar 'Top Down' mechanism · Bottom up mechanism · Polar vortex · NAM · SAM · Brewer-Dobson circulation

10.1 Solar Influence: 'Top Down'

The works of Kodera and Kuroda (2002), Haigh (1996, 1999), Haigh et al. (2005), Thompson et al. (2005) and Baldwin and Dunkerton (2001) are in favour of 'top-down' mechanism and discussed below.

10.1.1 Solar Influence: 'Top-Down' – via Polar Vortex and Lower Stratosphere

A possible mechanism was proposed by Kodera and Kuroda (2002) that suggests solar influence in the equatorial troposphere can be originated through changes in the meridional circulation via the tropical stratosphere. According to them, the strength of polar stratospheric jet is influenced by the solar heating anomalies. It subsequently impacts the path of upward travelling planetary waves. Those deposit their zonal momentum on the poleward side of the jet, causing weakening of the Brewer–Dobson circulation. It thus warms the lower tropical stratosphere in maximum of solar years compared to minimum years (as also shown in Fig. 3.7 and discussed in details in Sect. 3.8).

© Springer International Publishing AG, part of Springer Nature 2018

I. Roy, *Climate Variability and Sunspot Activity*, Springer Atmospheric Sciences, https://doi.org/10.1007/978-3-319-77107-6_10

10.1.2   Solar Influence: 'Top-Down' – via Lower Stratosphere to Troposphere

Perturbation in lower stratosphere has implication in the circulation fields of the troposphere is discussed in various studies by Haigh (1996, 1999, 2003), and Haigh et al. (2005), also shown in Fig. 9.1. Results for the influence of the 11-year solar cycle on the lower atmosphere were presented by Haigh (1999) using atmospheric circulation model. A pattern of response is found for high irradiance in which the midlatitude Ferrel cells and subtropical jets move poleward and the tropical Hadley cells broaden and weaken. The modelling study by Haigh (1996) also noted such a pattern. The changes in dynamics cause warming in subtropics and a characteristic band structure (vertical) of midlatitude variations of temperature. Despite the lack of a stratospheric polar vortex and the presence of a uniform stratosphere, Haigh et al. (2005) through simplified global circulation model (GCM) (without ocean) showed it is still possible to reproduce the patterns of troposphere.

10.1.3   Solar Influence: 'Top-Down' – via Stratospheric Polar Vortex to Polar Troposphere

Despite the sharply contrasting stationary wave climatologies and land–sea distributions in the southern and northern hemisphere, there are strong similarities between the meridional structures of the annular modes (Thompson and Wallace 2000) in two hemisphere. It supports some robust effects that lead from tops. Polar temperature around stratopause can also potentially affect polar lower atmosphere through the downwards propagation of polar annular modes as observed by Baldwin and Dunkerton (2001) (as shown in Fig 3.5 and discussed in details in Sects. 3.6 and 3.7). By presenting composites of time height development of SAM index, Thompson et al. (2005) showed that the same is also true for southern hemisphere polar region. By compositing difference between weak and strong stratospheric events, they suggested SAM follows similar mechanism (Fig. 10.1) as shown for NAM earlier.

10.2   Solar Influence: 'Bottom-Up'

In various studies, Meehl et al. (2003, 2008, 2009) proposed that solar effect in the troposphere is originated through ocean pathway and known as 'bottom-up' mechanism. Based on a modelling study, a new mechanism was suggested (Meehl et al. 2003) that is related to sea–air radiative coupling at the tropics. According to them, induced by cloud distributions, there are spatial asymmetries of solar forcing. It results in greater evaporation in the subtropics and subsequent transport of moisture

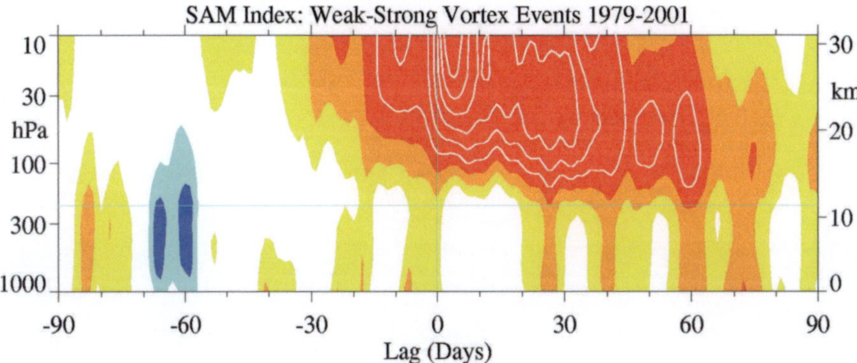

Fig. 10.1 Composite difference of the SAM index between weak and strong stratospheric events (Thompson et al, 2005, ©American Meteorological Society. Used with permission)

into the convergence zones of tropics. Through dynamically coupled atmosphere–ocean interaction, it produces higher precipitation. Meehl et al. (2004) also suggest that solar forcing produces coupled dynamical interactions in the tropics that strengthen the Walker and Hadley cell. The mechanism is also discussed by Meehl et al. (2008, 2009) which is illustrated in Fig. 10.2.

The global average solar forcing during DSO peaks compared to the minimum is $0.2\,\mathrm{Wm^{-2}}$ (Lean et al. 2005). At tropical latitudes, around the top of the atmosphere, it is 1–$2\,\mathrm{Wm^2}$. According to Meehl et al. (2008), the forcing can be considerably greater than the global average, where the Sun is most directly overhead, that is in the tropics. This extra energy is responsible to trigger coupled air–sea interactions. In relatively cloud-free areas of the subtropics, the increase of net solar heat energy input is translated into surface moisture and increased latent heat flux. It is then carried into the ITCZ and South Pacific Convergence Zone (SPCZ) to strengthen those features. That correspondingly intensifies the west-east Walker cell and consequently the south-north Hadley circulation. Thus, there are stronger trades that contribute to the reduction of SSTs in the subtropics and increased latent heat flux. In the tropical Pacific, the stronger trades are additionally allied with a dynamical response including enhanced upwelling which again contribute to a decrease in SSTs. There is also increased outflow in upper-level atmosphere from the SPCZ and ITCZ, and stronger subsidence is developed in the areas of subtropics. It produces consequently even lesser clouds and more radiation from the Sun making it to the surface (Fig. 10.2). Finally, it causes above the average pressure in the north Pacific over the places of Aleutian Low. Solar 'top-down' and 'bottom-up' mechanisms were also addressed in recent studies by Roy (2010, 2014) and Roy and Haigh (2012).

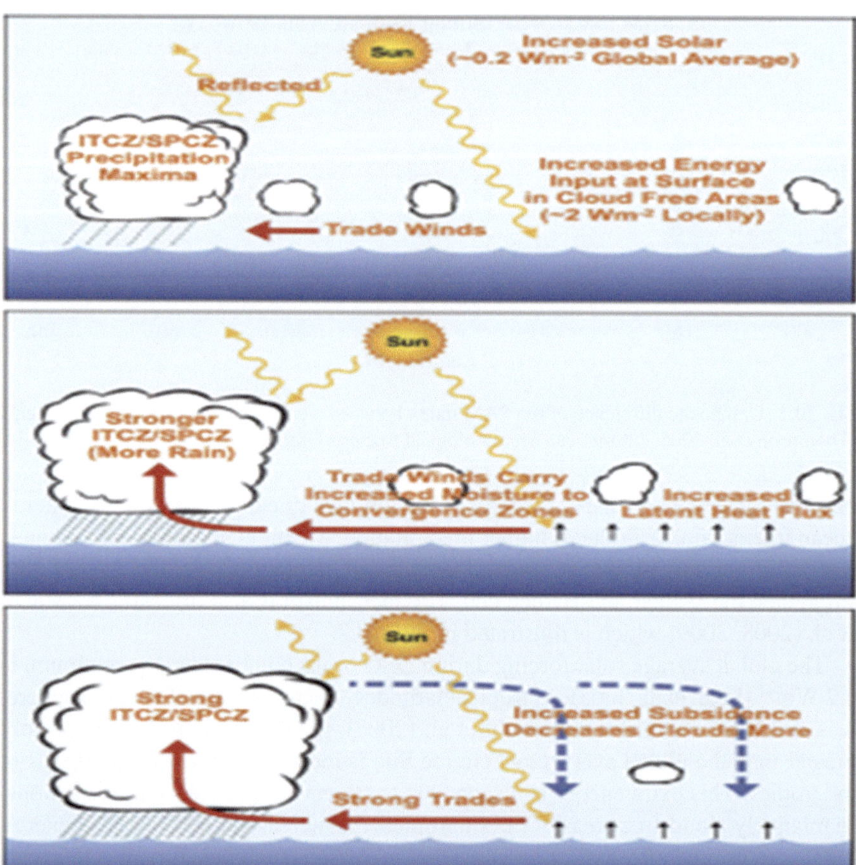

Fig. 10.2 Schematic showing proposed mechanism involved in the Pacific (N–S direction) during peak solar years. (Meehl et al. (2008), ©American Meteorological Society. Used with permission)

References

Baldwin MP, Dunkerton TJ (2001) Stratospheric harbingers of anomalous weather regimes. Science 294(5542):581–584. https://doi.org/10.1126/science.1063315

Haigh JD (1996) The impact of solar variability on climate. Science 272(5264):981–984

Haigh JD (1999) A GCM study of climate change in response to the 11-year solar cycle. Q J R Meteorol Soc 125(555):871–892

Haigh JD (2003) The effects of solar variability on the Earth's climate. Philos T R Soc A 361(1802):95–111

Haigh JD, Blackburn M, Day R (2005) The response of tropospheric circulation to perturbations in lower-stratospheric temperature. J Clim 18(17):3672–3685

Kodera K, Kuroda Y (2002) Dynamical response to the solar cycle. J Geophys Res 107(D24):4749. https://doi.org/10.1029/2002JD002224

Lean J et al (2005) SORCE contributios to new understanding of global change and solar variability. Sol Phys 230:27–53

Meehl GA, Washington WM, Wigley TML, Arblaster JM, Dai A (2003) Solar and greenhouse gas forcing and climate response in the 20th century. J Clim 16:426–444

Meehl GA et al (2004) Combinations of natural and anthropogenic forcings and 20th century climate. J Clim 17:3721–3727

Meehl GA, Arblaster JM, Branstator G, van Loon H (2008) A coupled air-sea response mechanism to solar forcing in the pacific region. J Clim 21(12):2883–2897

Meehl GA, Arblaster JM, Matthes K, Sassi F, van Loon H (2009) Amplifying the pacific climate system response to a small 11-Year solar cycle forcing. Science 325:1114–1118. https://doi.org/10.1126/science.117287

Roy I (2010) Solar signals in sea level pressure and sea surface temperature. Department of Space and Atmospheric Science, PhD Thesis, Imperial College, London

Roy I (2014) The role of the sun in atmosphere-ocean coupling. Int J Climatol 34(3):655–677. https://doi.org/10.1002/joc.3713

Roy I, Haigh JD (2012) Solar cycle signals in the Pacific and the issue of timings. J Atmos Sci 69(4):1446–1451

Thompson DWJ, Wallace JM (2000) Annular modes in the extratropical circulation. Part I: month-to-month variability. J Clim 13:1000–1016

Thompson DWJ, Baldwin MP, Solomon S (2005) Stratosphere-troposphere coupling in the Southern Hemisphere. J Atmos Sci 62:708–715

Chapter 11
An Overview of Solar Influence on Climate

Abstract An overview of the processes shown in determining the influence of the sun on climate is formulated in a holistic way and presented in the form of a flow chart, focusing on the Pacific region. It discussed hypotheses and evidence relating the combined influences of the Quasi-Biennial Oscillation, the El Niño–Southern Oscillation, and the solar cycle on the Walker and Hadley circulations in the context of ocean–atmosphere coupling. It suggests, in that coupling, that the Sun plays an important role, but it appears to be disturbed around the final half of the last century, probably related to climate change. This study leads towards a better understanding of the system of ocean–atmosphere coupling, accounting the variability of solar cyclic and will be useful for improved understanding of the climate–sun connection.

Keywords ENSO · QBO · Atmosphere-ocean coupling · Thermocline · Hadley cell · Walker cell · Ferrel cell · Solar 'Top-down' mechanism · 'Bottom-up' mechanism

11.1 Introduction

The sun is the prime driver of energy to the earth's climate. In terms of energy output, only 0.1% change from maximum to minimum years of 11-year cycle (Lean and Rind 2001), too negligible to influence the climate. But nowadays, there is a general consensus that the direct impact in the UV part of the spectrum changes (between 6 and 8% for solar minima to maxima) leads to more ozone and warming during solar maxima in the upper stratosphere (discussed in detail by Gray et al. 2010). Through the 'top-down' mechanism, solar signals can impact the lower stratosphere and troposphere and primarily driven from the upper stratosphere. The related mechanisms have been discussed by Haigh and co-workers (1996, 2005, 2006), Kodera and Kuroda (2002) and Baldwin and Dunkerton (2001). Meehl et al. (2008, 2009) posed the question of whether the effect of the sun in the troposphere is 'top-down' or 'bottom-up', i.e. forced by solar heating of the surface. To support both the views, different mechanisms were proposed, though a comprehensive understanding is still at large and highly required to entangle various discrepancies and reconcile controversies.

Based on 'top-down' mechanism, Haigh and co-workers (Haigh et al. 2005; Haigh 1996) showed, despite an absence of an ocean, without the stratospheric polar vortex

© Springer International Publishing AG, part of Springer Nature 2018

I. Roy, *Climate Variability and Sunspot Activity*, Springer Atmospheric Sciences, https://doi.org/10.1007/978-3-319-77107-6_11

and the presence of a uniform stratosphere, the simplified general circulation model (GCM) studies could observe a solar impact on tropospheric mean meridional circulation, characterising an expansion and weakening of the Hadley circulations, alongside a poleward movement of the Ferrel circulations (shown in Fig. 9.1). Using upper air data, Brönnimann et al. (2007a) also observed that with an increase in solar irradiance, there is a poleward movement of the Ferrel cell and subtropical jets (STJ). A possible mechanism was proposed by Kodera and Kuroda (2002) whereby the solar effect can impact equatorial stratosphere through changes in the meridional circulation that involve polar vortex (shown in Fig. 3.7). According to them, the solar heating anomalies can influence the path of upward propagating planetary waves via altering the strength of polar upper stratospheric jet. These waves warm the tropical lower stratosphere in solar maximum years by depositing their zonal momentum on the poleward side of the jet and subsequently weakening the Brewer–Dobson circulation (BDC). Intensification of the polar jet in upper tropospheric is also coupled with the positive phase of surface NAM (or Arctic oscillation (AO)) and surface SAM (or Antarctic oscillation (AAO)). Baldwin and Dunkerton (2001) in an observational analysis showed that perturbations in the polar vortex appear to be related to the downward propagation via the Southern annular modes and Northern annular modes (SAM and NAM). A dynamical mechanism was discussed that might communicate anomalies in stratosphere downward to the surface and troposphere via polar modes (shown in Fig. 3.5). There are strong resemblance between the meridional structures of the polar annular mode in the southern hemisphere (SH) and NH, and they follow similar formation mechanisms (Thompson and Wallace 2000) (shown in Fig. 10.1). Meehl and co-workers (2008, 2009) introduced and discussed 'bottom-up' mechanism. They explained a mechanism related to air–sea-radiative coupling, induced by cloud distributions, whereby the spatial asymmetries of sun's forcing cause greater evaporation in the subtropics and consequent transport of moisture into the tropical convergence zone. Thus around tropical Pacific it intensifies the trade winds (Fig. 10.2).

There is some observational evidence of solar influence on tropospheric climate. Christoforou and Hameed (1997) studied the variations of semi-permanent pressure systems, the Pacific High (PH) and the Aleutian Low (AL), and showed that variability in the sun impacts the location of their centre of actions. It thus causes large anomalies in regional climatic conditions and changes in storm tracks (shown in Fig. 5.3). AL also exhibits significant variations in intensity, apart from shifting its location. Van Loon et al. (2007) (vL07) and Meehl et al. (2008) using the technique of solar max compositing and Roy and Haigh (2010) using multiple linear regression technique (MLT) also confirmed this observation (shown in Fig. 5.1). Haigh (2003) suggests that there is a positive response of the sun in the lower tropical stratosphere which extends throughout the troposphere in a vertical band structure in both the hemispheres via midlatitudes (around 40–50°N with a maximum amplitude of 0.5°K) (shown in Fig. 9.2). Using the multiple linear regression technique of the ERA-40 dataset, Frame and Gray (2010) also noted a positive solar response (for the period 1979–2008), in the annual temperature, for both hemispheres, at midlatitudes of troposphere (shown in Fig. 5.8).

Studies indicate that, it is necessary to segregate the meteorological data according to the QBO phase, during the NH winter, to find a clear signature in the stratosphere for the solar 11-year cycle. Using data for the years 1956–1991, Labitzke and van Loon (1992), showed that warm upper stratospheric polar temperatures tend to occur during the QBO E-ly at solar minimum and QBO W-ly at solar maximum (shown in Fig. 3.4). It has been claimed that during El Niño winters, the polar vortex in northern stratosphere is warmer and perturbed to that from La Niña winters. It was also showed that ENSO-warm years are significantly warmer in the stratosphere during winter, at the northern hemisphere midlatitudes and poles, than the ENSO-cold years (Camp et al. 2007). The strength of NH wintertime stratospheric polar vortex is allied with the cold (warm) phase of ENSO (Thompson et al. 2002), and a pronounced strengthening (weakening) is also observed (shown in Fig. 4.1).

Using an EOF analysis of SST (data from 2nd half of the twentieth century), White et al. (1997) identified a leading variability mode whose spatial pattern suggests warming around the tropics of the Pacific, which is in phase with 11-year solar cycle (shown in Fig. 5.4). Data analysis (using period 1979–2000) observed that cold events of the ENSO are linked with dominant positive AAO and vice versa during the austral summer (DJF) (Carvalho et al. 2005). The alternation in phases of AAO was also shown to be associated with the intensity of polar jet (around 60 °S) and the latitudinal migration of upper level (200 hPa) STJ (around 45 °S) (shown in Fig. 4.2). Positive AAO phases are allied with the weakening and poleward shift of the subtropical characteristic, accompanied by an intensification of the high-latitude feature. Using multiple regression technique, Haigh and Roscoe (2006) (data from the last half of the last century) observed an anti-correlation between the ENSO and lower tropospheric polar modes. During the final half of the last century, climate change could be responsible for a weakening of both the tropical cells – the Walker and Hadley circulations. Weakening of the Walker cell in a warming climate was first documented by Held and Soden (2006). Vecchi and Soden (2007), using observational analyses, also indicated such weakening, during the last half of twentieth century, which is more pronounced in the Walker circulation. The two mechanisms 'bottom-up' coupled atmosphere–ocean surface response and the 'top-down' stratospheric response were investigated by Meehl et al. (2009), in versions of three global climate models. They compared the results with an observed solar signal that was detected using the method of compositing. According to them, it is much more appropriate to combine the mechanisms than the sum of their individual effects.

This study discusses how the sun, QBO and ENSO all play important parts in regulating tropospheric climate. It shows that for solar influence on the troposphere, both the proposed mechanisms, i.e. forcing from the below (with an SST influence) and forcing from the top (stratospherically driven) are operative, although the responses are likely to be influenced by the background state of the atmospheric. Due to climate change during the latter half of the twentieth century, the ocean–atmosphere coupling appears to be disturbed.

11.1.1   Methodology

The current work proposes a comprehensive overview including both the ocean (mainly involving the Pacific Ocean) and atmosphere, to account for the solar influences. It uses results of data analysis, and this work is supported by evidences from other highly cited popular research.

In terms of data analysis, it primarily applies the technique of multiple regression as also been used in recent studies (Roy and Collins 2015; Roy et al. 2016; Roy 2018). The main advantage of this technique is that it can segregate other influences that might contaminate/influence the signal of the sun. Here, the codes of multiple regression with an order one autoregressive noise model (AR(1)) developed by Myles Allen (personal communication, University of Oxford, UK) are used. Multiple linear regression (MLR) may be written as

$$y = \beta X + u$$

where 'X' is a 'n x m' order matrix, which are thought to influence the data and comprise time series of m indices. 'y' is a rank n vector that contains the time series of the data. 'β' is a rank m vector containing amplitudes of the indices, which we are going to estimate. The noise term is represented by 'u', which may arise due to various sources (e g. unmodelled variability, all sources of observational error, internal noise, etc.). Using an autoregressive noise model (AR(1)) of order one, autocorrelation in the time series and its effects on the derived coefficients of regression are estimated alongside its significance levels. Finally, the level of confidence in the value of derived β is estimated for each index, using the Student's t-test.

Multiple regression analysis of sea surface temperature (SST) and sea level pressure (SLP) data is performed during 1850–2005. The independent indices used are a linear trend, optical depth (OD), ENSO, solar cycle variability and often QBO. To represent longer-term climate change, the linear trend is used. As the focus of this study is on variability in solar 11-year cycle, there is a little effect on the choice of longer-term trend on the derived signal.

To represent solar cycle variability monthly sunspot number (SSN) is mainly used. Two other total solar irradiance (TSI) datasets are also used; apart from SSN, Foster's TSI is primarily based on observation on SSNs, whereas the Solanki and Krivova dataset is mainly based on relative SSN and facular brightening. Over the past century, volcanic aerosols have been playing an important part for forcing in global climate. Hence, in this analysis, stratospheric aerosol optical depth (OD) has been considered as an independent factor. We used the Niño 3.4 index for ENSO, which is defined as the 3-month running mean of SST departures, about the 1971–2000 base period, in the Niño 3.4 region (120–170°W, 5°N-5°S). Time series of independent variables are shown in Fig. 11.1. QBO data from period 1953 onward are available (between levels 90 and 3 hPa). Recent QBO reconstructions that extend back to 1900 are also available now. (See all data sources at the end)

Here, a flow chart is formulated in steps, depicting a consolidated overview of ocean–atmosphere coupling, supported by mechanisms and observations. In formulating this flowchart, results from the regression analyses have been incorporated. The three major climate variabilities are shown with oval outlines, which are solar, QBO and ENSO. The major circulations appear by non-rectangular parallelograms, which are responsible for modulating the effect of the main variabilities. The pathways of the signals start from 'A', initiated by solar variability with all other signals marked by labels ('A'–'Z'). The direction of change in behaviour during the steps is shown by '-' (for decrease) and '+' (for increase). Subscripts indicate steps of the

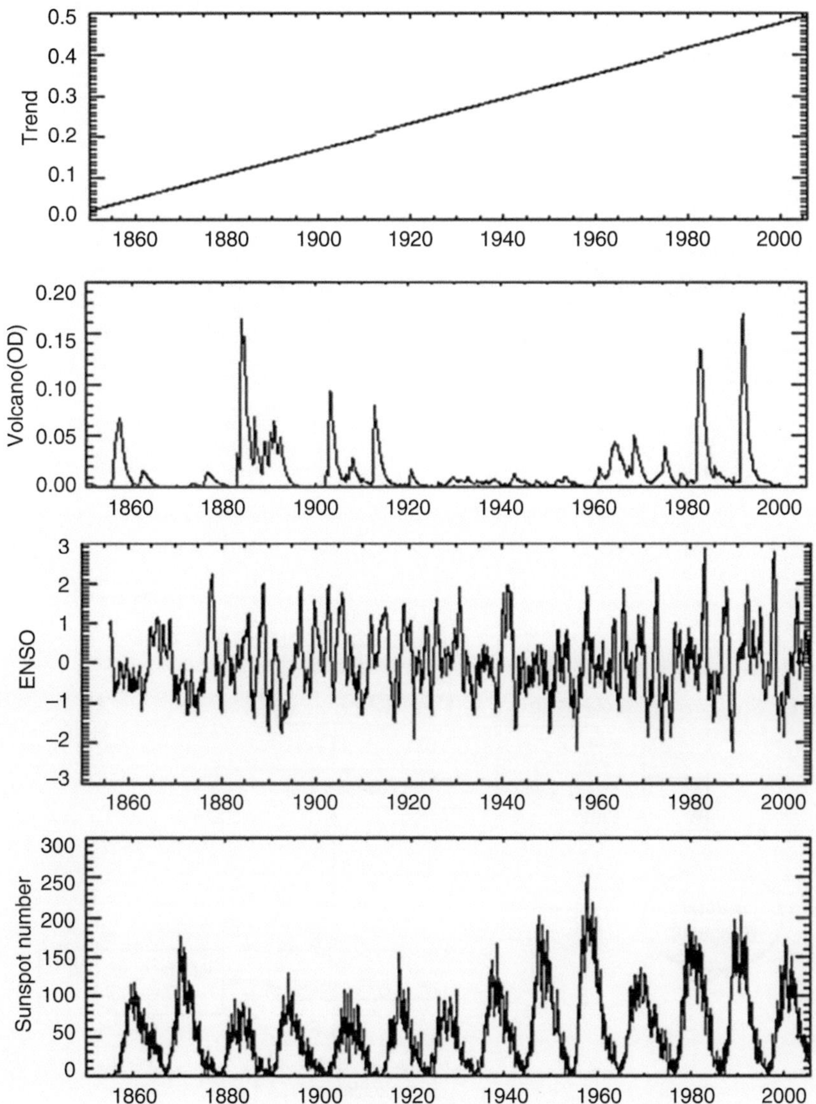

Fig. 11.1 Time series of trend, volcano, ENSO and sunspot number

same process, and superscripts suggest same effect but different forcing – radiative or dynamical. All acronyms used in the flow chart is presented at the end.

11.2   Representative Results: Figure and Tables

All figures and tables related to representative results are presented in this section (Figs. 11.2 to 11.17 and Table 11.1). Relevant texts are discussed in Sect. 11.3.

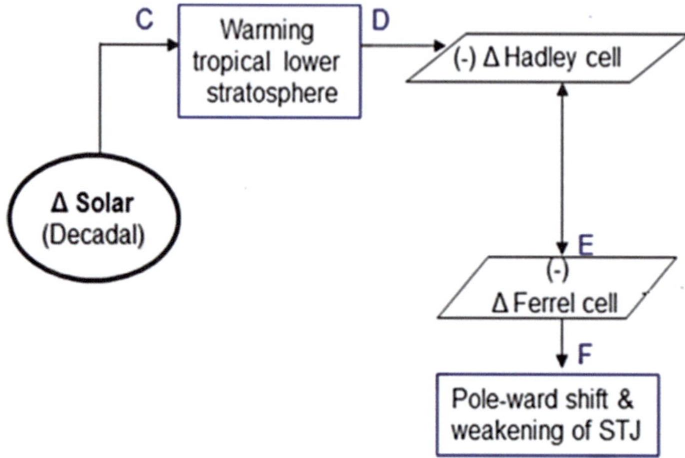

Fig. 11.2 Flow chart for solar influence (atmosphere only) via lower stratosphere

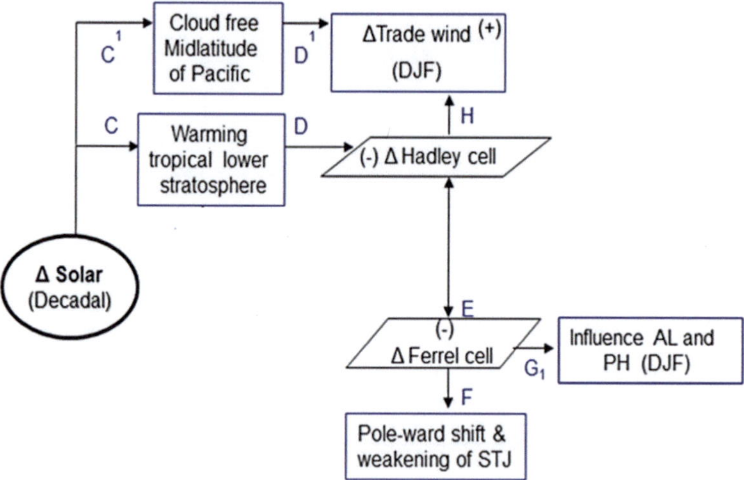

Fig. 11.3 Flow chart for solar influence (only atmosphere) via midlatitude of the Pacific and lower stratosphere

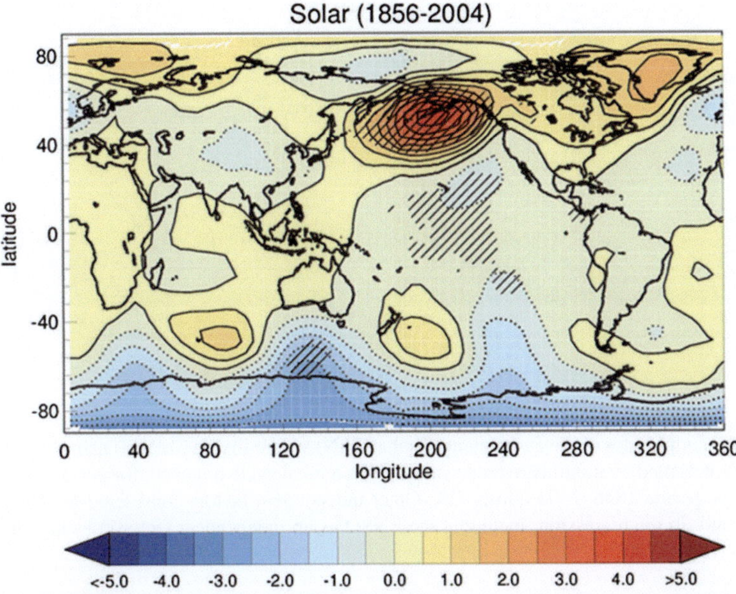

Fig. 11.4 Amplitudes of the components of the solar (using monthly SSN) variability of SLP (in hPa) during DJF. Other independent factors are OD, trend and ENSO. Hatching indicates areas assessed statistically significant at the 5% level, and dashed lines indicate negative values (Roy 2014)

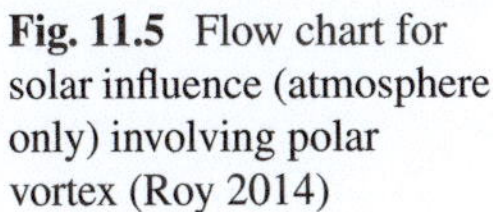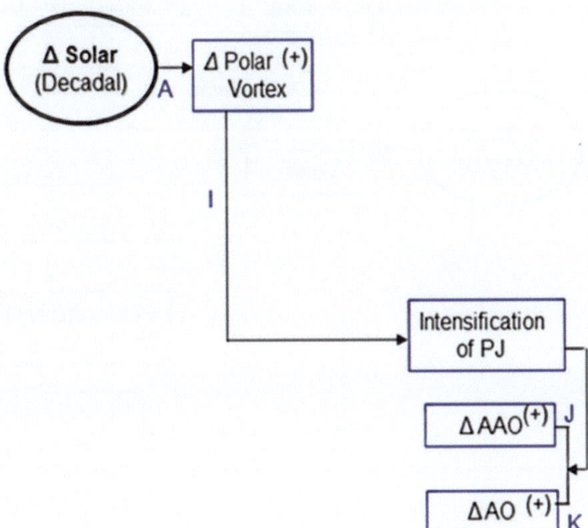

Fig. 11.5 Flow chart for solar influence (atmosphere only) involving polar vortex (Roy 2014)

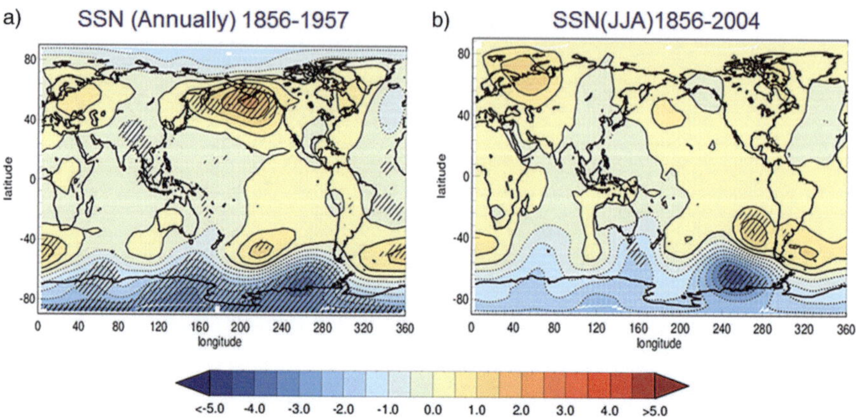

Fig. 11.6 Amplitudes of the component of solar (SSN) variability on SLP: (**a**) using all months of a year, represented by monthly average values, with annual cycle removed, for period 1856–1957; (**b**) for the period 1856–2004 during JJA. Other independent factors considered are the OD and linear trend. In the regression, including the ENSO as an independent factor does not alter these findings (Roy 2014)

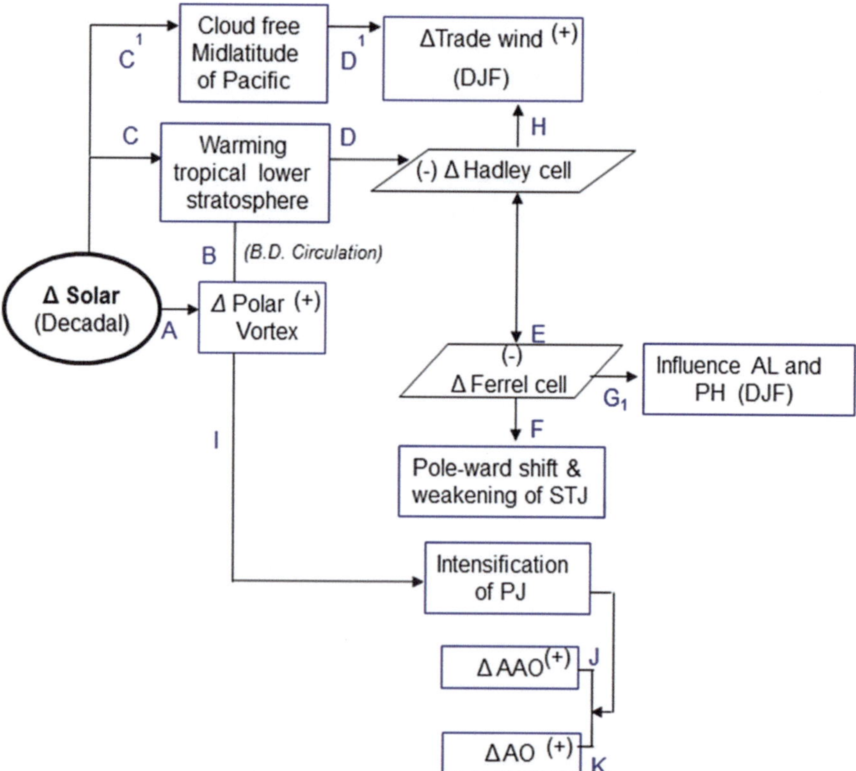

Fig. 11.7 Flow chart for solar influence (atmosphere only) involving polar vortex and lower stratosphere (Roy 2014)

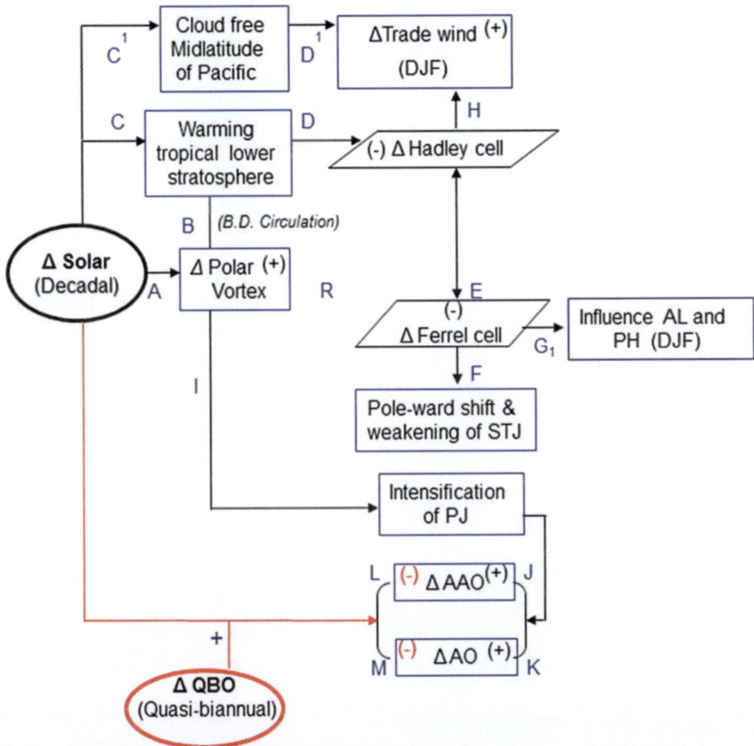

Fig. 11.8 Flow chart for stratosphere–troposphere coupling involving Sun and QBO (Roy 2014)

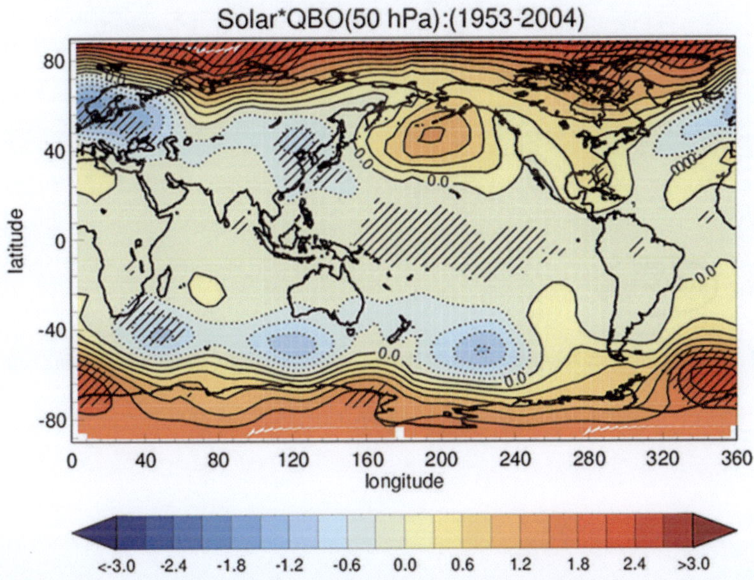

Fig. 11.9 Results of multiple linear regression technique (1953–2004) of SLP data. The components (hPa) due to the compound signal for solar*QBO (at 50 hPa) is presented. Other independent factors used are the OD, trend and ENSO, (Roy 2014)

Fig. 11.10 Flow chart for ocean coupling (only Pacific) (Roy 2014)

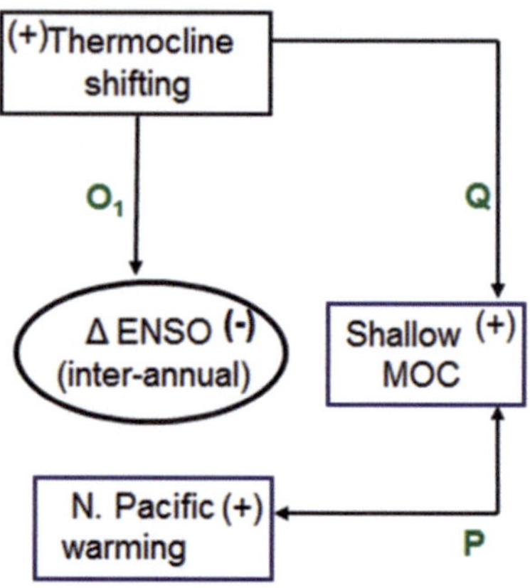

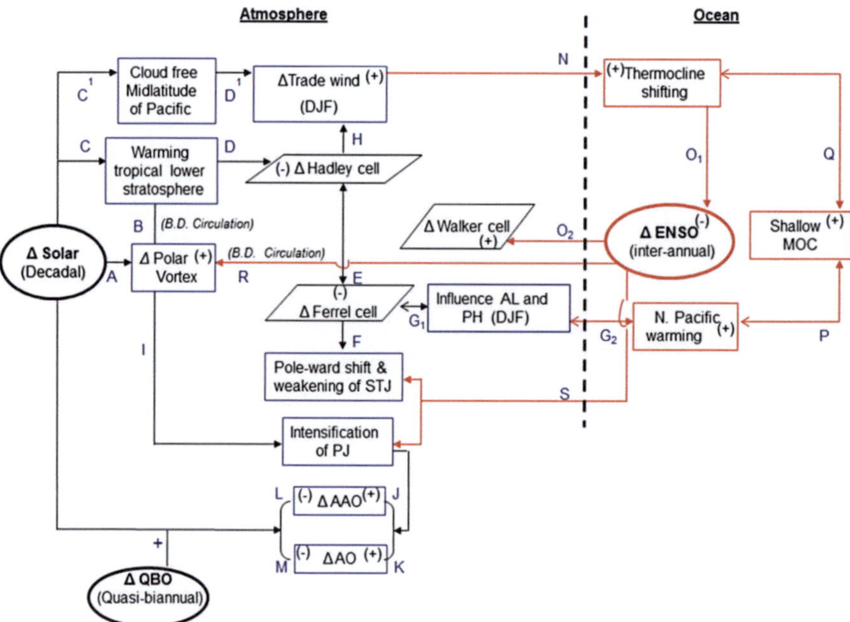

Fig. 11.11 Flow chart showing ocean (only Pacific) and atmosphere coupling involving sun, QBO and ENSO (Roy 2010, 2014)

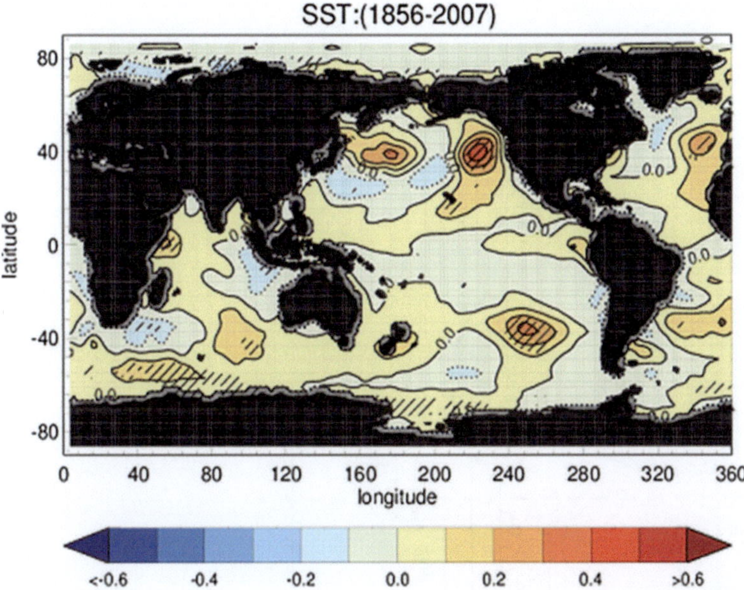

Fig. 11.12 Amplitudes of the components of the solar variability (using monthly SSN) of SST (in °K). Other independent parameters used are OD, trend and ENSO. All months of a year are considered, (represented by monthly average values), from NOAA dataset, with annual cycle removed (Roy 2014)

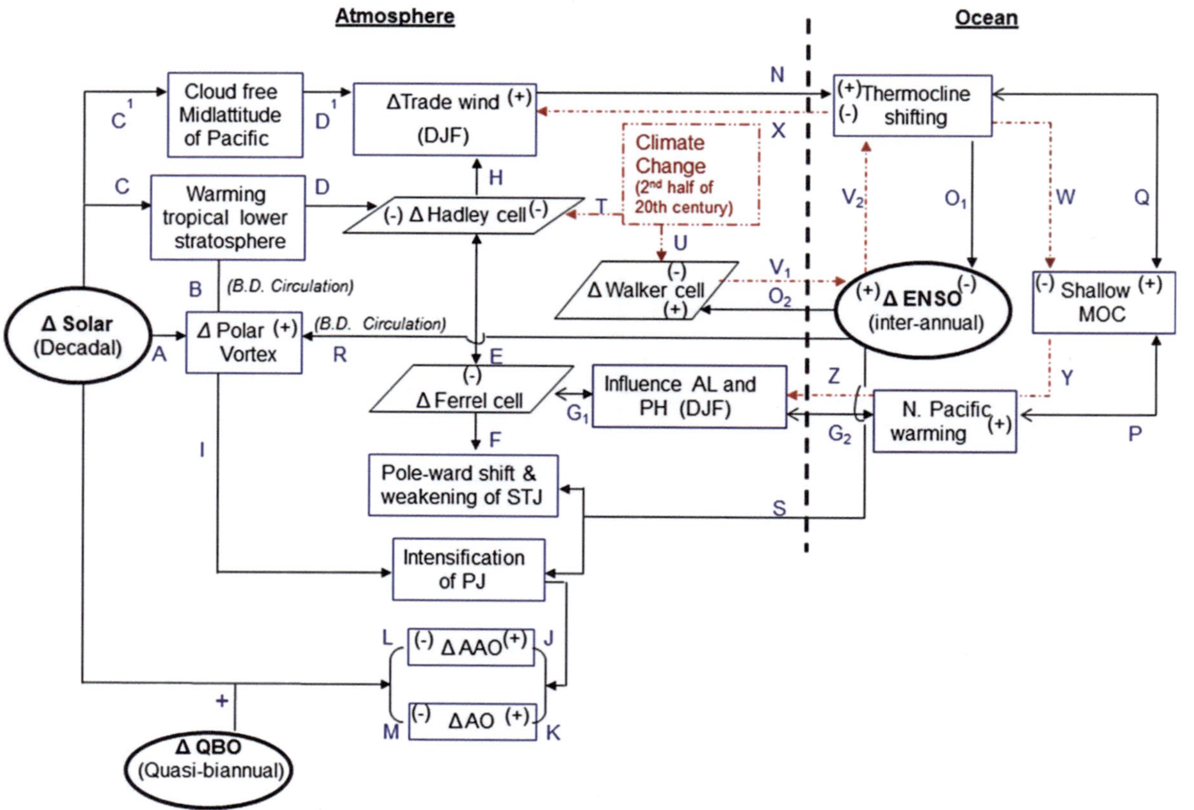

Fig. 11.13 Flow chart showing ocean (only Pacific) and atmosphere coupling involving sun, QBO, ENSO and climate change (Roy 2014)

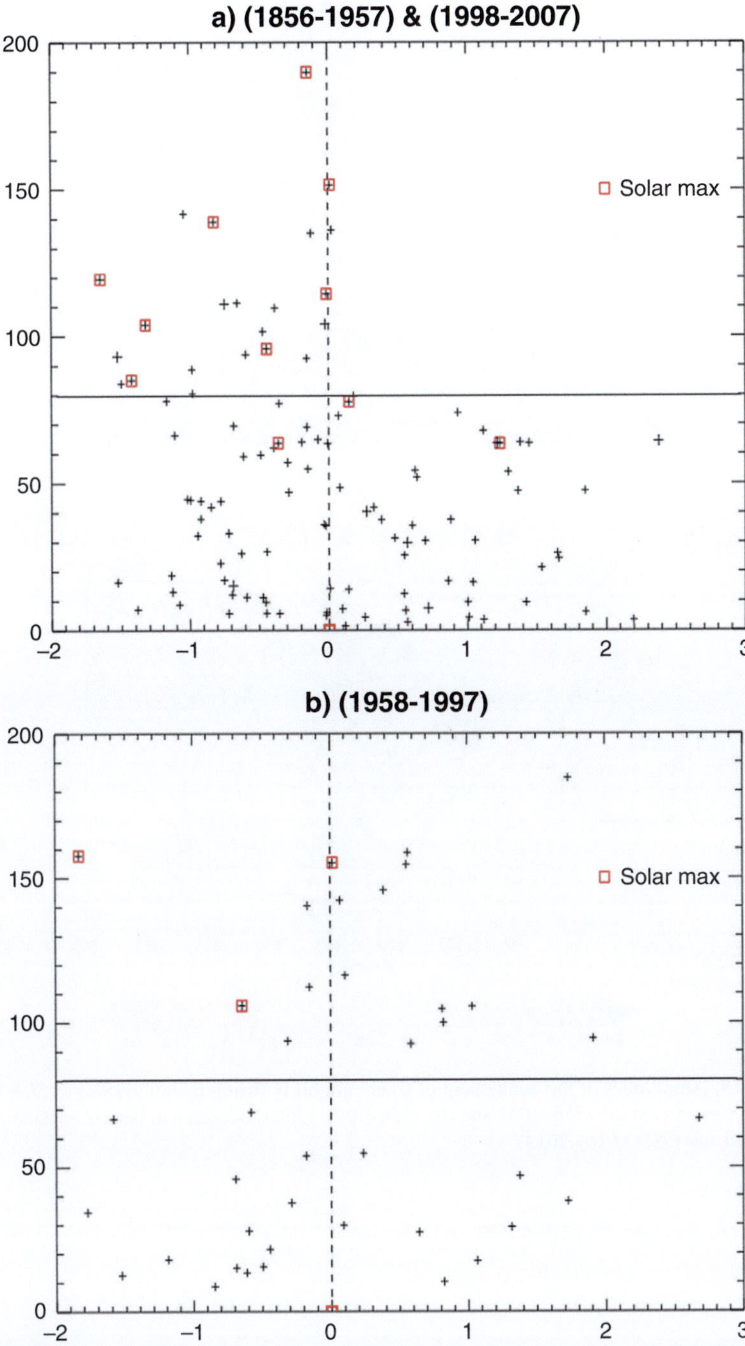

Fig. 11.14 Scatter diagram of annual mean SSN vs DJF means ENSO index during: (**a**) (1856–1957) and (1998–2007); (**b**) (1958–1997) (Roy 2014)

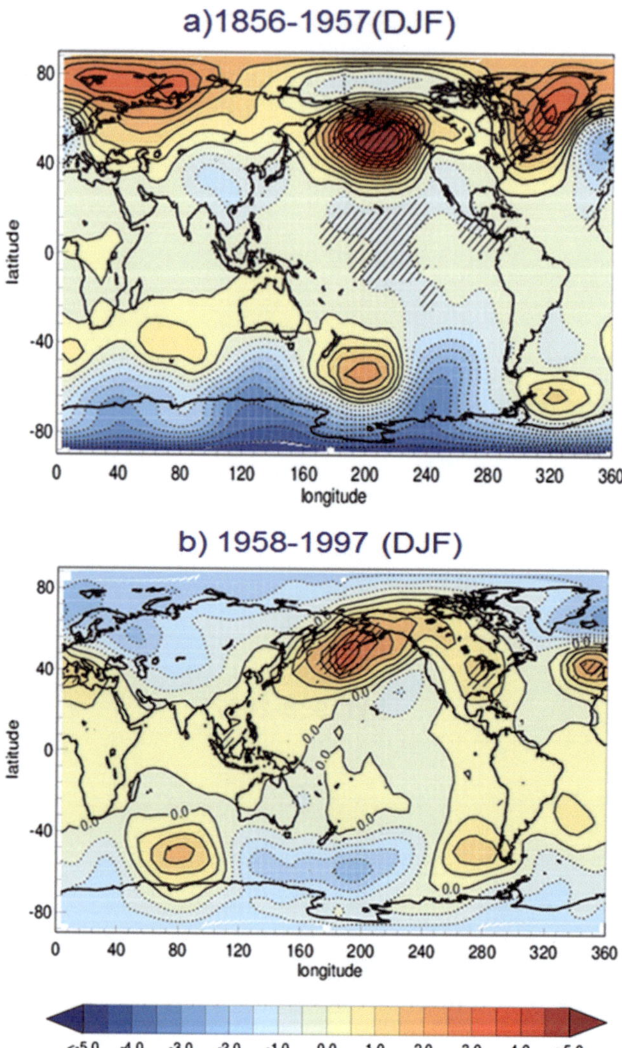

Fig. 11.15 Amplitudes of the component of solar variability (using monthly SSN) of SLP (DJF) in hPa for the period (**a**) 1856–1957 and (**b**) 1958–1997. Other independent factors considered are OD, trend and ENSO (Roy 2014)

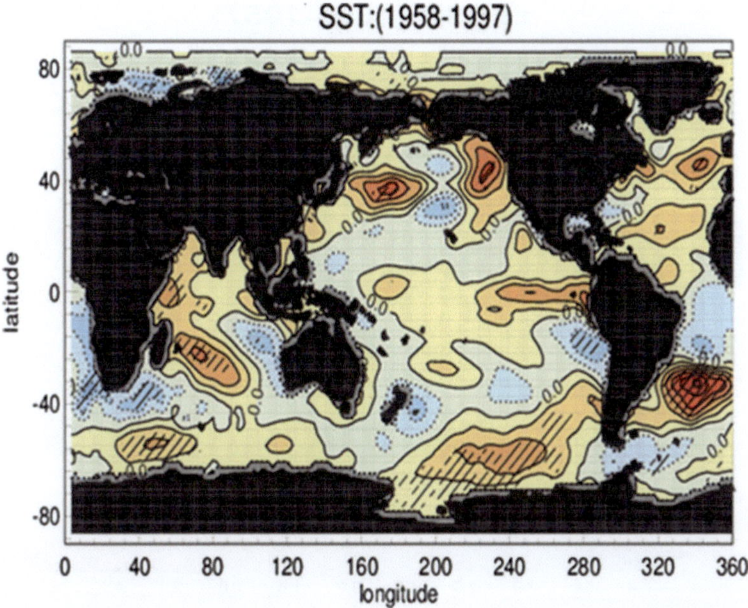

Fig. 11.16 Amplitudes of the components of the solar variability (using monthly SSN) of SST (1958–1997). Regression was done annually removing annual cycle, and other independent factors are trend and OD (Roy 2014)

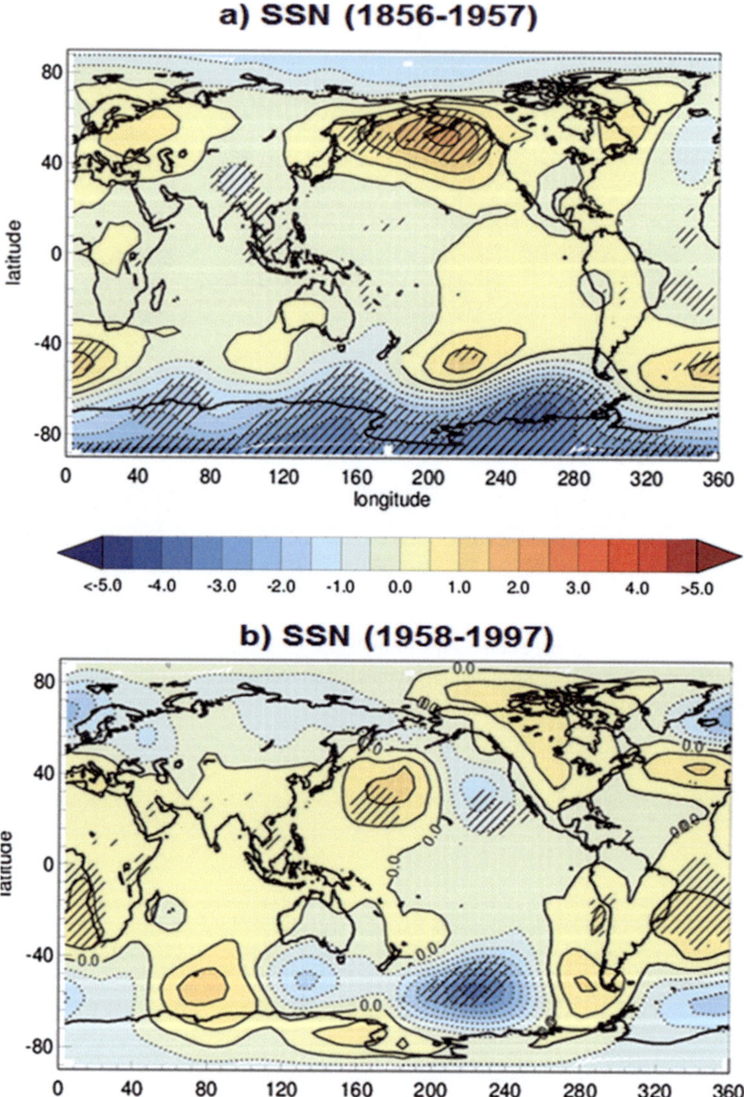

Fig. 11.17 (**a, b**): Amplitudes of the component of solar (SSN) variability of SLP: (**a**) for the period 1856–1957 and (**b**) for the period 1958–1997. Regression was done annually with removal of annual cycle. Other independent factors are trend and OD (Roy 2014)

Table 11.1 To indicate whether the pathways are hypothesised or evidenced (Roy 2014)

			Mechanism	
		Observation	Explained	Hypothesised
Atmosphere (A–M)	Lower stratosphere and midlatitude of Pacific (A–H)	A,B,D,E,F,G,H	A,B,C,D,E,F,H	G,H
	Upper stratosphere (A, I–M)	A,I,J,K,L,M	A	
Ocean–atmosphere coupling (N–S)		O,P,Q,R,S,G	R,N,O	P,G
Climate change (T–Z)		T,U,V,W,X,Y,Z	T,U,V,X	Z

11.3   Results Text

11.3.1   Atmosphere Only: Sun and QBO

Pathways showing atmosphere only is marked by 'A–M' with following steps: sun via lower stratosphere by 'C–F'; cloud-free midlatitude 'G–H'; polar vortex 'A,I–K'; and QBO by 'L,M'. B connects lower stratosphere and the polar vortex. Pathways are mentioned below with appropriate references.

Atmosphere only: the sun via *lower stratosphere* (Fig. 11.2, C–F)

C:	Haigh (1996), Frame and Gray (2010)
D, E, F:	Haigh (1996, 1999), Haigh et al. (2005), Brönnimann et al. (2007a)

Atmosphere only: the sun via *midlatitude of Pacific and lower stratosphere* (Fig. 11.3, C–H)

C^1, D^1:	Meehl et al. (2008, 2009)
G:	Van Loon et al. (2007), Christoforou and Hameed (1997), Fig. 11.4
H, G:	Fig. 11.4, Meehl et al. (2008, 2009), van Loon et al. (2007)

Atmosphere only: the sun through upper stratospheric region (via polar modes) (Fig. 11.5: A, I–K)

A, I:	Usual mechanism following thermal wind balance
J–K:	Baldwin and Dunkerton (2001), Fig. 11.6.

Atmosphere only: the sun via *cloud-free midlatitude, lower stratosphere and the polar vortex* (Fig. 11.7, **A–K**)

B:	Kodera and Kuroda (2002).

Atmosphere only: the Sun in combination with QBO (Fig. 11.8, A–M)

A flow chart, showing how the atmosphere is influenced by solar cycle variability that involves cloud-free midlatitude, polar vortex and lower stratosphere, was marked by (A–K). Here the additional pathways due to QBO are also included and shown by (L–M) with red.

L–M   Fig. 11.9.

11.3.2   Ocean (Only Pacific) and Atmosphere Coupling: Sun, QBO and ENSO

The coupling only in the Pacific Ocean is shown in Fig. 11.10. It is combined with Fig. 11.8 that involves the sun and QBO in the atmosphere (demonstrated by 'A' to 'M') to formulate Fig. 11.11. In Fig. 11.11, (A–S) the additional pathways due to ENSO that considers atmosphere and ocean together are included and marked by (N–S) with red.

N, O$_2$	Usual ENSO mechanism
O$_1$	Roy and Haigh (2010, 2012)
Q and **P**	Zhang et al. (2006), Veechi et al. (2007), Mantua and Hare (2002)
G$_2$	Fig. 11.12
R	Thompson et al. (2002), Camp and Tung (2007)
S, J, K	Carvalho et al. (2005), Haigh and Roscoe (2006)

11.3.3   Atmosphere and Ocean (Only Pacific) Coupling: Sun, QBO, ENSO and Climate Change

Atmosphere and ocean coupling involving sun, QBO and ENSO were shown by 'A' to 'S' in Fig. 11.11. Here in Fig. 11.13, the additional pathways due to climate change are also included and shown by (T–Z) with red dotted dash.

U, T, W:	Vecchi and Soden (2007), Held and Soden (2006), McPhaden and Zhang (2004)
V$_1$, V$_2$:	Usual ENSO mechanism, Fig. 11.14
X, Z, V$_1$:	Roy and Haigh (2012), Figs. 11.15a, b and Fig. 11.16
Y:	Fig. 11.16

11.4   Discussion

It is a holistic representation, in a form of a flow chart, showing how the solar cycle variability influences the atmosphere and ocean (mainly, Pacific). Figure 11.1 shows the time series of various independent parameters used in the MLR. Figure 11.2

(C–F) shows how the sun via lower stratosphere influences Hadley and Ferrel cells, considering 'top-down' mechanism. In active phase of a solar cycle, the sun warms tropical lower stratosphere (Frame and Gray 2010). It subsequently can cause weakening of Hadley and Ferrel cell with weakening and poleward shift of STJ (Haigh 1999, 1996), Haigh et al. (2005), (Bronnimann et al. 2007a). Figure 11.3. includes midlatitude of Pacific and involves 'bottom-up' mechanism as discussed by Meehl et al. (2008, 2009). Here the sun causes intensification of the trade wind around the tropical Pacific as marked by paths C^1 and D^1. To represent solar radiative forcing through the region of cloud-free midlatitude of the Pacific that acts alongside dynamical forcing of the sun, superscripts are introduced and shown by C and D. As also demonstrated by observational results, both of these forcings have potential to increase the strength of the Pacific trade wind (Fig. 11.4). Several observational studies (van Loon et al. (2007), Christoforou and Hameed (1997), Fig. 11.4) detected the influence on AL and PH as shown by G in the flow chart of Fig. 11.3. Figure 11.5 demonstrates solar effect via upper stratosphere, which is initiated with 'A' and well explained by usual mechanism following thermal wind balance. How the influence from polar vortex is transported down to troposphere in a form of polar annular modes is supported by the work of Thompson and Wallace (2000) and Baldwin and Dunkerton (2001). It is also in agreement with the results of MLR as shown in Fig. 11.6. Figure 11.7 illustrates the connection between lower stratosphere and polar vortex which follows mechanism proposed by Kodera and Kuroda (2002). Figure 11.8 included the influence of QBO and subsequent changes in polar annular modes. A new index combining SSN and QBO is constructed following the results of Baldwin and Dunkerton (2001) and Labitzke and van Loon (1992) and used in MLR. It suggests active sun-westerly QBO and less active sun-easterly QBO both triggers negative AO and AAO features (Fig. 11.9). The coupling only in the Pacific Ocean is shown in Fig. 11.10 that includes ENSO, thermocline shifting in east Pacific and tropical Pacific shallow meridional overturning circulation (MOC). Figure 11.11 shows pathways including ENSO that involves both the atmosphere (Fig. 11.8) and the ocean (Fig. 11.10). It follows usual ENSO mechanism and another study (Mantua and Hare 2002). Influence of the ENSO around polar vortex as shown by 'R' in Fig. 11.11 follows Camp and Tung (2007) and Thompson et al. (2002). Whereas, Haigh and Roscoe (2006) and Carvalho et al. (2005) suggest there are influence of ENSO on polar annular mode. Figure 11.12 shows north Pacific warming as detected in MLR study. In Fig. 11.13, the additional pathways due to climate change are also included. It followed general ENSO mechanism and combined the work of Vecchi and Soden (2007) and McPhaden and Zhang (2004), which suggested a substantial decrease in the strength of tropical Pacific shallow MOC after the 1950s with modest intensification since 1998. The similar feature is also noticed for Walker and Hadley circulation. Figure 11.15 suggests there is no solar signature around the eastern Pacific of ITCZ during a later period, and signal around AL also weakened. Warming in tropical E. Pacific after 1950s is noticed in MLR analysis in Fig. 11.16 that follows White et al. (1997). However, when ENSO is considered in the regression that warming disappears. Climate change also affects signals in polar annular modes as shown in Fig. 11.17.

Table 11.1 suggests how each link is explained by a mechanism, evidenced by observation, or only be viewed at this stage as a hypothesis. The purpose of the flow chart is mainly to capture behaviour of the sun during low and high phases of its 11-year cycle and thus, not only restrict to few solar peak years from the 11-year cycle (as considered by Meehl and co-authors (2007, 2008, 2009) in compositing methods). Apart from SSN as representative of solar cycle variability, total solar irradiance (TSI) reconstruction from Krivova and Solanki (2003) and Foster (2004) is also used, but the primary results suggest similar for solar cycle variability using either TSI or SSN (hence not shown). It shows both the proposed solar mechanisms on the troposphere, i.e. forcing from the below (with an SST influence) and from the top (stratospherically driven) are operative, although the stratospheric and tropospheric dynamic responses are influenced by the background state of atmosphere. It indicates that during the last half of the twentieth century, both the 'bottom-up' and 'top-down' routes designate differently and show how the coupling is disturbed during time period influenced by climate change.

This current analysis still suggests few areas which are unresolved and mentioned here (a–c). a) The interaction among major modes of variability: solar, ENSO and QBO act together, and it is still unclear about the true nature of the linearity of their interaction. b) The mechanism how the sun impacts on SSTs: during high solar (HS) years (say SSN above 80), the sun has been shown to impact tropical SSTs before 1958 and after 1997 but at the lower solar period to be overwhelmed by the inherent strong ENSO variability (Fig. 11.14a, b). How the apparent influence is reflected/communicated in a coupled ocean–atmosphere system is still not fully understood. c) Climate change during the period 1950–1997 probably induced the ocean–atmosphere coupled system to act in a different way, and this also affected the solar signature (Fig. 11.17a, b). Such observations identify the requirement for quantifying the actual signature of the sun without and with the effect of climate change. Hence, into the task of characterisation of solar influence, it introduces more complications and indicates why actual quantification of solar influence is so difficult.

The pathways as proposed and presented in the flow chart (Fig. 11.13) might lead towards a better understanding of ocean–atmosphere coupled system, taking into account cyclic variability of the sun and will be useful for improved understanding of the climate–sun connection.

Acronyms

AAO:	Antarctic oscillation
AL:	Aleutian low
AO:	Arctic oscillation
BDC:	Brewer–Dobson circulation
DJF:	December–January–February
ENSO:	El Niño and southern oscillation
GCM:	General circulation model
HADSLP:	Hadley centre sea level pressure
HADSST:	Hadley centre sea surface temperature
HS:	Higher solar
ITCZ:	Intertropical convergence zone

JJA:	June–July–August
LS:	Lower solar
MLR:	Multiple linear regression
MOC:	Meridional overturning circulation
NAM:	Northern annular mode
NAO:	North Atlantic oscillation
NCEP:	National centres for environmental prediction
NH:	Northern hemisphere
NOAA:	National oceanic and atmospheric administration
OD:	Optical depth
PH:	Pacific high
QBO:	Quasi-biennial oscillation
SAM:	Southern annular mode
SH:	Southern hemisphere
SLP:	Sea level pressure
SSN:	Sunspot number
STJ:	Subtropical jet
SST:	Sea surface temperature
TSI:	Total solar irradiance

Data Source

- For **SLP**, HadSLP2 dataset from http://www.hadobs.org is used (Allan et al. 2006).
- For **SST**, HadSST2 dataset from http://hadobs.metoffice.com/hadsst2/ is used (Rayner et al. 2006).
- For **SSN**, data used from ftp://ftp.ngdc.noaa.gov/STP/SOLAR_DATA/ SUNSPOT_NUMBERS/MONTHLY .
- **TSI** datasets used: the Solanki and Krivova dataset (Solanki and Krivova 2003) and Foster's dataset (Foster 2004).
- **OD** data used are available from http://data.giss.nasa.gov/modelforce/strataer/ tau_line.txt (Sato et al. 1993).
- For **ENSO**, Niño 3.4 index, obtained from http://climexp.knmi.nl, is used.
- **QBO** data, available from http://www.pa.op.dlr.de/CCMVal/Forcings/qbo_data_ ccmval/u_profile_195301-00412.html and QBO reconstructions from Brönnimann et al. (2007b).

Summary In Part II, it covered about atmosphere and ocean coupling and the role of the sun in regulating the coupling. The initial focus was on the ocean and it described various relevant definitions. It then defines the solar's so-called 'top-down' and 'bottom-up' mechanism and discusses a few related research. Next, a general overview of solar influence on climate in a form of flow chart is presented. It discusses how initiated by solar eleven-year cyclic variability, the overall atmosphere-ocean coupling takes place. Those also involve QBO, ENSO, polar annular modes and climate change. It mainly shows why it is so difficult to detect robust solar influence on the climate of the Earth and indicates about complications.

In the final part (III), it will discuss various other main influences on climate. It will compare periods of different phases of PDO and AMO and will discuss about Arctic and Antarctic snow cover. Some results from CMIP5 model outputs will be presented and compared with observation. There are also discussions on greenhouse effect, major volcanos and Antarctic ozone hole. Other solar outputs are also important in modulating the climate of the earth and will be discussed briefly in the final chapter.

References

Allan R, Ansell T (2006) A new globally complete monthly historical gridded mean sea level pressure dataset (HadSLP2): 1850–2004. J Clim 19:5816–5842. https://doi.org/10.1175/JCLI3937.1

Baldwin MP, Dunkerton TJ (2001) Stratospheric harbingers of anomalous weather regimes. Science 294(5542):581–584. https://doi.org/10.1126/science.1063315

Brönnimann S, Annis JL, Vogler C, Jones PD (2007a) Reconstructing the quasi-biennial oscillation back to the early 1900s. Geophys Res Lett 34:L22805. https://doi.org/10.1029/2007GL031354

Brönnimann S, Ewen T, Griesser T, Jenne R (2007b) Multidecadal signal of solar variability in the upper troposphere during the 20th century. Space Sci Rev 125 (1–4):305–317

Camp CD, Tung KK (2007) The influence of the solar cycle and QBO on the late-winter stratospheric polar vortex. J Atmos Sci 64(4):1268–1283. https://doi.org/10.1175/JAS3883.1

Camp CD et al (2007) Stratospheric polar warming by ENSO in winter: A statistical study. Geophys Res Lett 34:L04809. https://doi.org/10.1029/2006GL028521

Carvalho LMV, Jones C, Ambrizzi T (2005) Opposite phases of the Antarctic Oscillation and relationships with intraseasonal to interannual activity in the tropics during the austral summer. J Clim 18:702–718. https://doi.org/10.1175/JCLI-3284.1

Christoforou P, Hameed S (1997) Solar cycle and the Pacific 'centers of action'. Geophys Res Lett 24(3):293–296. https://doi.org/10.1029/97GL00017

Foster S (2004) Reconstruction of solar irradiance variations, for use in studies of global climate change: application of recent SoHO observations with historic data from the Greenwich Observatory, PhD thesis, University of Southampton, Southampton

Frame THA, Gray LJ (2010) The 11-year solar cycle in ERA-40 data: an update to 2008. J Clim 23:2213–2222. https://doi.org/10.1175/2009JCLI3150.1

Gray LJ et al (2010) Solar influences on climate. Rev Geophys 48:RG4001. https://doi.org/10.1029/2009RG000282

Haigh JD (1996) The impact of solar variability on climate. Science 272(5264):981–984

Haigh JD (1999) A GCM study of climate change in response to the 11-year solar cycle. Q J Roy Meteor Soc 125(555):871–892

Haigh JD (2003) The effects of solar variability on the Earth's climate. Philos T R So A 361(1802):95–111

Haigh JD, Roscoe HK (2006) Solar influences on polar modes of variability. Meteorol Z 15(3):371–378. https://doi.org/10.1127/0941-2948/2006/0123

Haigh JD, Blackburn M, Day R (2005) The response of tropospheric circulation to perturbations in lower-stratospheric temperature. J Clim 18(17):3672–3685

Held IM, Soden BJ (2006) Robust responses of the hydrological cycle to global warming. J Clim 19:5686–5699

Kodera K, Kuroda Y (2002) Dynamical response to the solar cycle. J Geophys Res 107(D24):4749. https://doi.org/10.1029/2002JD002224

Labitzke K, van Loon H (1992) On the association between the QBO and the extratropical stratosphere. J Atmos Terr Phys 54(11/12):1453–1463

Lean J, Rind D (2001) Earth's response to a variable Sun. Science 292(5515):234–236

Mantua NJ, Hare SR (2002) The Pacific decadal oscillation. J Oceanogr 58(1):35–44

McPhaden MJ, Zhang D (2004) Pacific Ocean circulation rebounds. Geophys Res Lett 31:L18301. https://doi.org/10.1029/2004GL020727

Meehl GA et al (2007) Global climate projections. In: Solomon S et al (eds) Climate change 2007: the scientific basis: contribution of working group I to the fourth assessment report of the inter-governmental panel on climate change. Cambridge University Press, Cambridge, pp 747–845

Meehl GA, Arblaster JM, Branstator G, van Loon H (2008) A coupled air-sea response mechanism to solar forcing in the Pacific region. J Clim 21(12):2883–2897

Meehl GA, Arblaster JM, Matthes K, Sassi F, van Loon H (2009) Amplifying the Pacific Climate System response to a small 11-year solar cycle forcing. Science 325:1114–1118. https://doi.org/10.1126/science.117287

Rayner NA, Brohan P, Parker DE, Folland CK et al (2006) Improved analyses of changes and uncertainties in sea surface temperature measured in situ since the mid-nineteenth century: The HadSST2 dataset. J Clim 19(3):446–469

Roy I (2010) Solar signals in sea level pressure and sea surface temperature. Department of Space and Atmospheric Science, PhD Thesis, Imperial College, London

Roy I (2014) The role of the sun in atmosphere-ocean coupling. Int J Climatol https://doi.org/10.1002/joc.3713

Roy I (2018) Solar cyclic variability can modulate winter Arctic climate. Sci Rep 8(1):4864. https://doi.org/10.1038/s41598-018-22854-0

Roy I, Haigh JD (2010) Solar cycle signals in sea level pressure and sea surface temperature. Atmos Chem Phys 10(6):3147–3153

Roy I, Haigh JD (2012) Solar cycle signals in the Pacific and the issue of timings. J Atmos Sci 69(4):1446–1451. https://doi.org/10.1175/JAS-D-11-0277.1

Roy I, Collins M (2015) On identifying the role of Sun and the El Niño Southern Oscillation on Indian Summer Monsoon rainfall. Atmos Sci Lett 16(2):162–169

Roy I, Asikainen T, Maliniemi V, Mursula K, (2016) Comparing the influence of sunspot activity and geomagnetic activity on winter surface climate. J Atmos Sol Terr Phys 149:167–179

Sato M, Hansen JE, McCormick MP, Pollack JB (1993) Stratospheric aerosol optical depths (1850–1990). J Geophys Res 98(22):987–994

Solanki SK, Krivova NA (2003) Can solar variability explain global warming since 1970? J Geophys Res 108(A5):1200. https://doi.org/10.1029/2002JA009753

Thompson DWJ, Wallace JM (2000) Annular modes in the extratropical circulation. Part I: month-to-month variability. J Clim 13:1000–1016

Thompson DW, Baldwin MP, Wallace JM (2002) Stratospheric connection to Northern Hemisphere Wintertime weather: implications for prediction. J Clim 15:1421–1428. https://doi.org/10.1175/1520-0442. 015<1421:SCTNHW>2.0.CO;2

van Loon H, Meehl GA, Shea DJ (2007) Coupled air-sea response to solar forcing in the Pacific region during Northern Winter. J Geophys Res 112(D2):D02108

Vecchi GA, Soden BJ (2007) Global warming and the weakening of the tropical circulation. J Clim 20(17):4316–4340

White WB, Lean J, Cayan DR et al (1997) Response of global upper ocean temperature to changing solar irradiance. J Geophys Res Oceans 102(C2):3255–3266

Zhang D, McPhaden MJ (2006) Decadal variability of the shallow Pacific meridional overturning circulation: relation to tropical sea surface temperatures in observations and climate change models. Ocean Model 15(3–4):250–273

Part III
Other Major Influences on Climate

Abstract In the final part, we discuss the other main influences on climate. Chapter 12 elaborates about possible limitations in detecting and explaining robust solar influences. Here major discussions were based on the method of solar peak year compositing and about the sun-NAO connection. We explained it is important that associated limitations are recognised and corrected so that theories proposed to describe any solar influence on climate are based on robust evidence. Afterwards, we compared periods of various phases of PDO and AMO. Chapter 13 covers discussions about Arctic and Antarctic snow cover. The sharp decline in ice around Arctic is noticed. Antarctic is accumulating ice in past few decades, and there is a rising trend. West Antarctica, however, shows a dramatic decrease in ice cover. In Chap. 14, we presented results from CMIP5 model outputs and compared with observation. Some results from historical, as well as RCP scenario, were discussed. CMIP5 models are also segregated considering Low or High Top and related teleconnections are compared with observation and AMIP5. Chapter 15 discusses on Green House Effect. Then in Chap. 16, we described years of major Volcanos and various effects on climate. Antarctic Ozone Hole and banning of CFCs through Montreal Protocol were also discussed, and it constituted Chap. 17. It was shown that ozone in the stratosphere and greenhouse gas in the troposphere both have a comparable effect on surface climate. Apart from solar eleven-year cyclic variability, as discussed here using SSN, other solar outputs are also important in modulating the climate of the earth and discussed briefly in Chap. 18.

Keywords Arctic and Antarctic ice · CMIP5 and AMIP5 models · Historical and RCP scenario · Green House effect · Volcanos · Ozone hole

Chapter 12
Sun: Atmosphere-Ocean Coupling – Possible Limitations

Abstract This chapter covers possible limitations in relevant analyses of atmosphere-ocean coupling initiated by the Sun. First, it focused on one study that discussed solar 'top-down' vs. 'bottom-up' mechanism. Then it addressed some limitations related to methodology and discussed solar peak year compositing method. Solar NAO connection as captured by multiple linear regression was also attended. It also discussed the solar lag relationship. At the end, years of various phases of PDO and AMO are compared, and their relevance was discussed relating to global climate.

Keywords Solar 'top down' influence · Bottom-up influence · NAO · PDO · AMO · ISM · Multiple Linear Regression Technique · Method of Peak year Compositing

12.1 Sun: Atmosphere-Ocean Coupling 'Top-Down' vs. 'Bottom-Up' Mechanism: a Case Study

The study of Meehl et al. (2009) is chosen to explore that further. Meehl et al. (2009) investigated two mechanisms: the 'bottom-up'-coupled atmosphere-ocean surface response and the 'top-down' stratospheric response. They showed a combination of both mechanisms 'bottom-up' as well as 'top-down' give a closest match to observation. They compared model results with their observations using versions of three global climate models (Whole Atmosphere Community Climate Model (WACCM)-fixed-SST, WACCM-coupled, Community Climate System Model version 3 (CCSM3)), with mechanisms acting together or alone. The CCSM3 only includes coupled air-sea mechanism, the 'bottom-up' route, with coupled components of atmosphere, land, ocean and sea ice but without interactive ozone chemistry or a resolved stratosphere. Whereas, a version of the WACCM only takes account of the 'top-down' mechanism. Here, the model is run with climatological SSTs, and solar variability is changed, while other external forcings are kept constant. It has no dynamically coupled air-sea interaction but resolved the stratosphere with fully interactive ozone chemistry. The observed pattern of temperature was not reproduced by either of these two models, on their own in a time scale of decadal solar

I. Roy, *Climate Variability and Sunspot Activity*, Springer Atmospheric Sciences, https://doi.org/10.1007/978-3-319-77107-6_12

scale. However, when a new hybrid model was developed, it produced a much closer result to the observations in the Pacific that used the atmospheric component from WACCM while coupled with the dynamical land, ocean and ice modules from CCSM3. Their results suggest each mechanism acting alone can indicate a weaker signal. However, if the two mechanisms act together, a response in the tropical Pacific is noticed that closely matches with their observations. It indicates that it is much more appropriate to consider the combination of mechanisms rather than the sum of their individual influences. Figure 12.1 shows results of their analysis for SST and Fig. 12.2 for precipitation. The observation that uses the method of peak/maximum solar compositing is shown on the top panel of these two plots. The bottom panel shows the combined results using WACCM-coupled model. However, there are several issues regarding the validation of such findings.

First of all, as indicated by IPCC (2007), their 'bottom-up' model is reported to generate frequent ENSO. Hence such a tendency may dominate after few runs. Moreover, making intercomparison of model results is also difficult as not all experiments are present in all models.

Moreover, their observed solar signals, which are compared with models, are also questionable. They did the compositing analysis for peak solar years, considering averages for DJF, with climatology subtracted – which is similar to that of vL07. However, using the compositing technique, the results of the detected signal may not only be sensitive to the choices of details of the method (Roy and Haigh 2012; Roy 2014: Tung and Zhou 2010). Those are also biased strongly due to mixing up the signal (Roy and Haigh 2010) with ENSO-like strong variability. As the peak SSN years are usually associated with the ENSO cold phase and thus using the technique of max year compositing (as used by vL07 and Meehl et al. 2009), the ENSO signal might be misinterpreted as a solar in the tropics. Meehl et al.'s (2009) results do indicate, however, the importance of atmosphere-ocean coupling that needs to be considered cautiously.

12.2 Sun: Atmosphere-Ocean Coupling – Limitations of Peak Year Compositing

12.2.1 Solar Cycle Signals in Peak Year Compositing for SLP: a Case Study

The study of van Loon and Meehl (2011) is considered in this regard. Figure 12.3 suggests the signal around South Pacific as noticed by van Loon and Meehl (2011) is sensitive to years chosen for composites. Figure 12.3a shows the signature deduced from differencing data from the 11 peak SSN years from cycles 12 to 22 (for solar cycle years, see Chap. 8, Table 8.1) with those of the climatology period 1950 to 1979, as done by van Loon et al. (2007). Comparing with that work suggests a negative area around 70–100° W, 70–80° N, a positive region south of the equatorial Pacific, a weak negative region in the north Pacific subtropics and a similar

Fig. 12.1 SST – showing a combination of mechanisms is much more appropriated. Composite for peak solar years (DJF) for observed SSTs (°C) is shown on top. The second panel shows results for average of five ensemble members for CCSM3 for the last century (°C). The third panel shows for model WACCM with specified non-varying SSTs, for 10 peak solar years. The last panel shows results for WACCM for 11 peak solar years (°C), which is coupled to the dynamical sea ice components, ocean and land from CCSM3. Significant at the 5% level is shown by stripe, while dashed lines mark locations of maxima for climatological precipitation. (From Meehl et al. (2009). Reprinted with permission from AAAS)

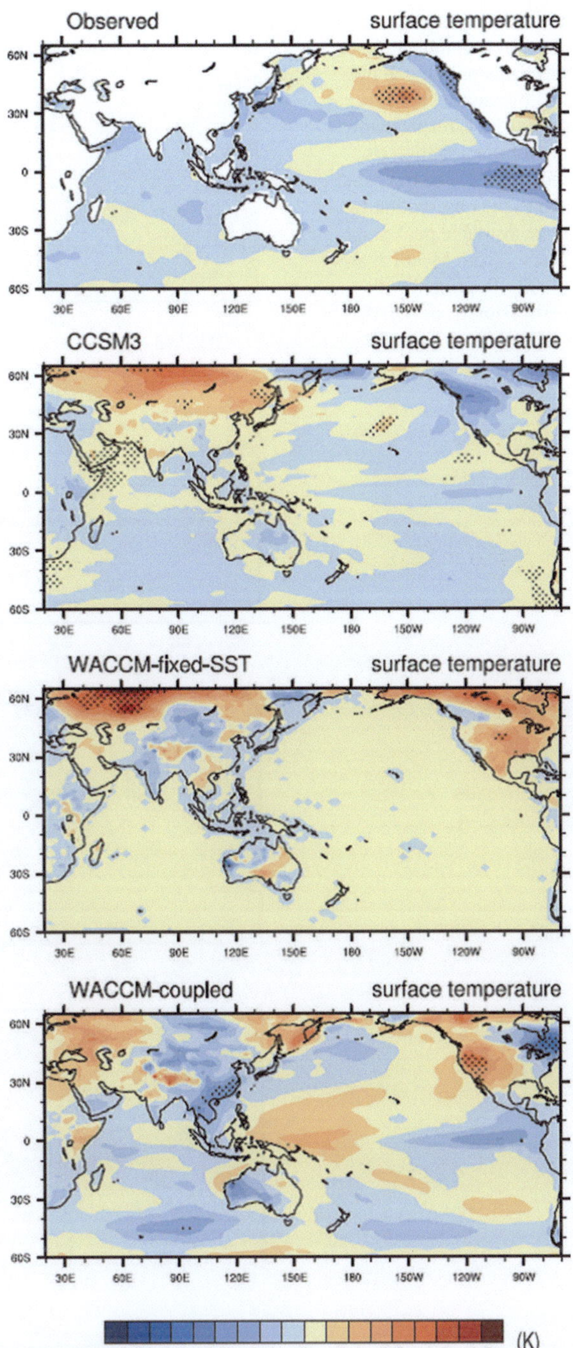

Fig. 12.2 Precipitation – showing a combination of mechanisms is much more appropriated. It is the same as Fig. 12.1, respectively, but for precipitation (in mm day $^{-1}$). The top panel shows the observation for three available peak solar years. (From Meehl et al. (2009). Reprinted with permission from AAAS)

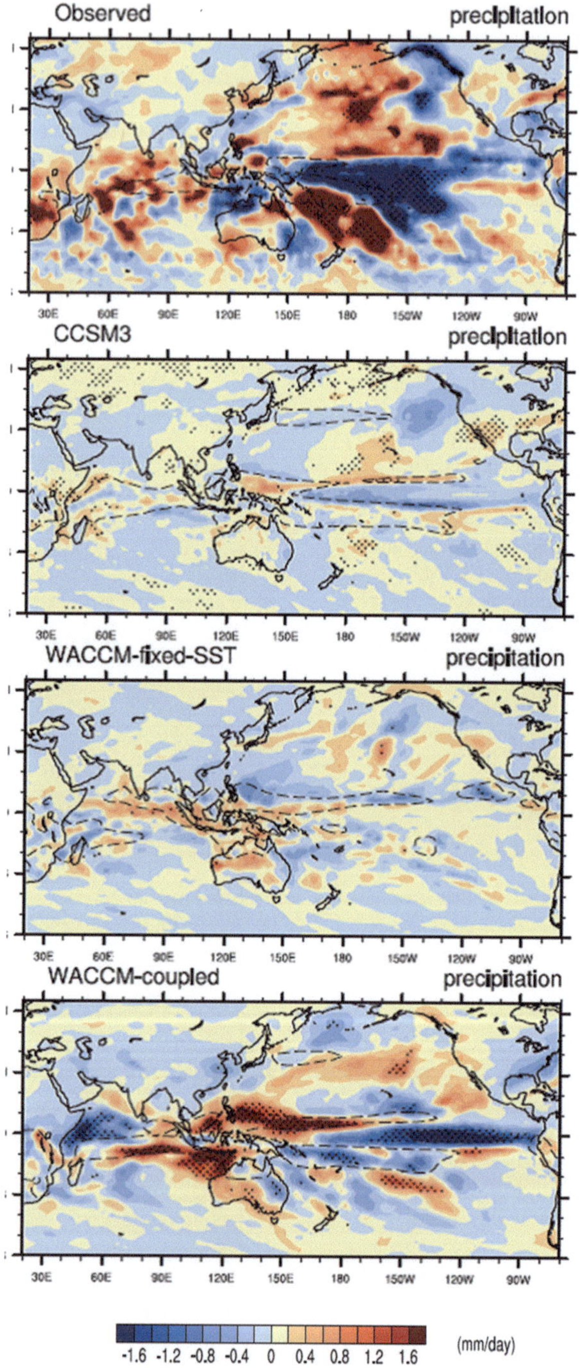

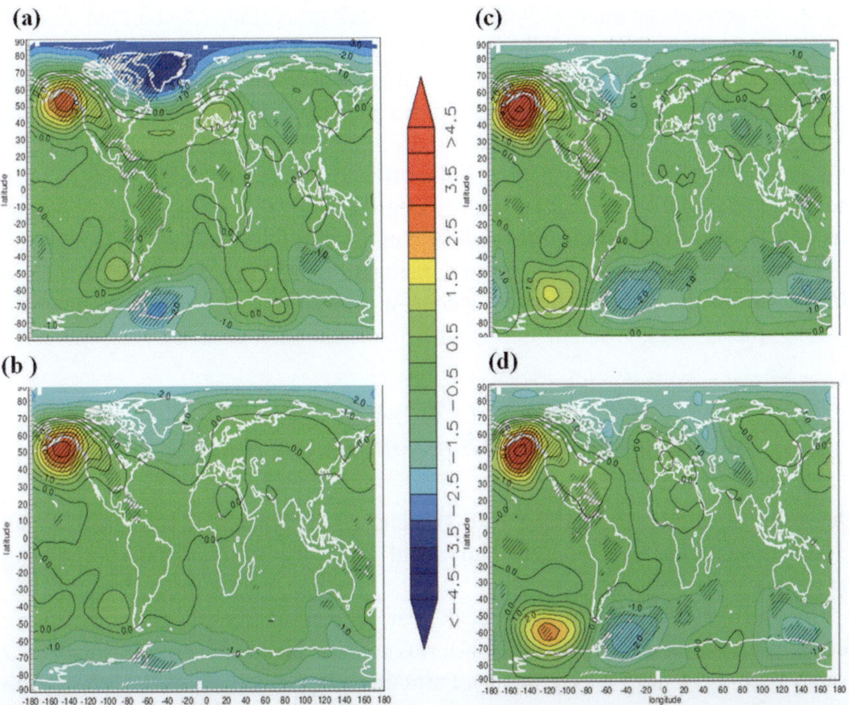

Fig. 12.3 The solar signal in the SLP (hPa) for HadSLP2 dataset following the method of solar peak year compositing (Jan–Feb). (**a**) Using 11 peak solar years during 1871 to 1989 about a climatology during 1950 to 1979 (12–22 cycle w.r.t. 1950–1979 avg) [the same as van Loon et al. (2007)]. (**b**) As in (**a**), but about the climatological period of 1871 to 1998 (12–22 cycle w.r.t. 1871–1998 avg) [similar to Meehl et al. (2008)]. (**c**) As in (a), but about a climatology of 1968 to 1996 (12–22 cycle w.r.t. 1968–1996 avg). (**d**) As in (**c**), but with 10 peak solar years during 1905 to 2000 (14–23 cycle w.r.t. 1968–1996 avg) [the same as van Loon and Meehl (2011)]. Dotted lines show negative contours, and regions significant at the 95% level using a t-test are marked by shading. (After Roy and Haigh 2012 ©American Meteorological Society. Used with permission)

strong positive signal in the region of the AL. The signals with higher magnitude, as noted by van Loon et al. (2007), may have resulted due to HadSLP1 dataset, an older version of the data. But these perhaps could be more likely due to a seasonal anomaly introduced by the use of DJF for the climatology but JF for peak years. Figure 12.3b considers a climatological period during 1871–1998 but uses similar solar cycle peak years; it is the same as used by Meehl et al. (2008). The firm signal is still observed in the north Pacific, but a weaker signal is noticed in almost all regions due to the use of this longer-period climatology. There is an exception perhaps in the equatorial west Pacific, particularly for SO seesaw, near its west end. For Fig. 12.3a, signals at high latitude, as identified by van Loon et al. (2007), particularly the negative patch over northeastern Canada and Greenland, are significantly weaker. The last two plots are restricting the climatology to more recent years (Fig. 12.3c, d). Figure 12.3d uses a climatology during 1968–1996 and takes the ten most recent solar peak years, as considered by van Loon and Meehl (2011), while

Fig. 12.3c presents an analysis with the same peak years (Fig. 12.3a–c) but uses the recent climatology (Fig. 12.3d). A signal in southern midlatitudes, particularly a positive anomaly around the 60° S, 120° W, has been produced only using more recent solar peaks, which is not noticed in the analysis of the longer/earlier datasets. In the equator the small dipole signal, as identified by van Loon et al. (2007) which is necessary for driving the apparent solar effect, has changed in strength and also moved. It again indicates that signatures identified by solar compositing technique are very sensitive to the detail methods of compositing.

Theories proposed to explain any solar signature in SLP, or more commonly on climate, should be based on firm evidence – it is necessary that this is recognized.

12.2.2   Solar Cycle Signals in Peak Year Compositing for Indian Summer Monsoon: a Case Study

The study of van Loon and Meehl (2012) is chosen to explore that further. To detect a solar signal in the ISM, van Loon and Meehl (2012) use 14 solar peak years during the period 1850–2004.

To investigate further, the solar peak year, as well as trough year compositing during the 155-year period (1850–2004), was considered that covered 14 solar peak/trough years (Fig. 12.5 a, b, c and d). Particular emphasis is on the regions of the Indian subcontinent. To test the sensitivity of the choices of solar years, trough year compositing is conducted alongside solar peak year compositing as shown in Fig. 12.4. In determining minimum years, a similar technique as used by vLM12 is

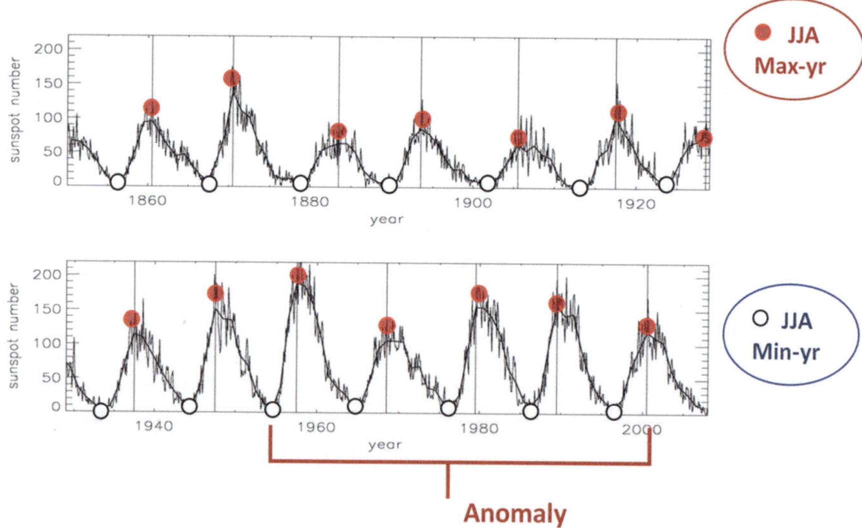

Fig. 12.4 Annual mean SSN (thick curve) and monthly mean (thin curve). Vertical lines with red spot designate the year of peak annual SSN for each solar cycle, while a blank spot for the year of minimum SSN

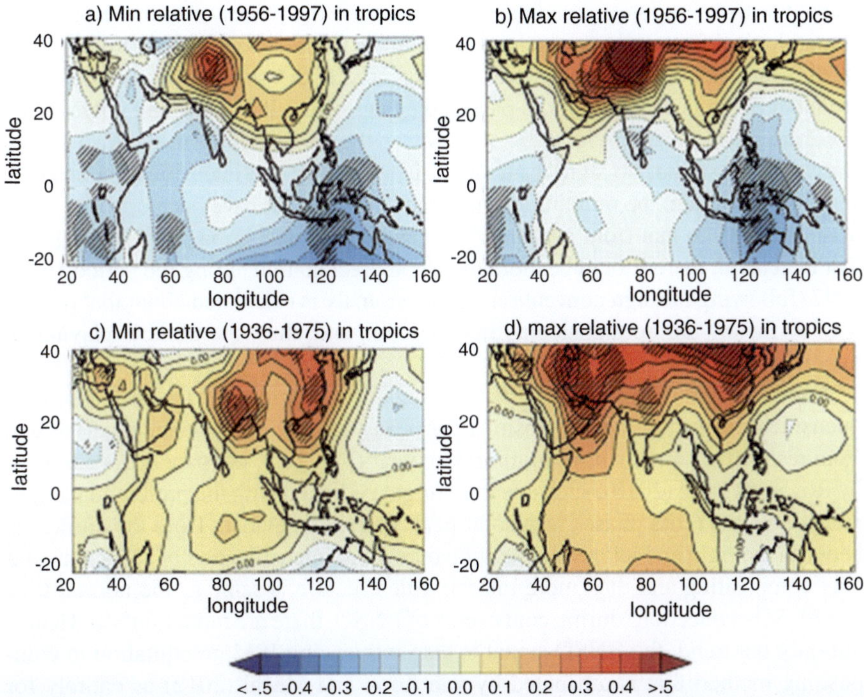

Fig. 12.5 (**a–d**): Solar signal using SSN peak/minimum year compositing. Focusing on Indian subcontinent in June-July-August (JJA) for HADSLP2 data (hPa), it considered 14 solar cycles during 1850 to 2004. In (**a**) the composite is for solar minimum years, and the climatology from 1956–1997 is subtracted from the period 1850–2004. In (**b**) the same climatology period is used but it is for solar peak year compositing. (**c**) and (**d**) are same as (**a**) and (**b**), respectively, but the climatology period removed is 1936–1975. Dotted lines show negative contours. Regions significant at the 95% level using a t-test are shaded (Roy and Collins 2015)

used, and the same solar peak years were employed. The left-hand panel of Fig. 12.5 uses the minimum (trough) year compositing, whereas the right panel shows a result for maximum (peak) year compositing. The period 1950s–1997s was identified as the period of weakening of both the Hadley and Walker cell but more in the Walker cell than the Hadley (e.g. Vecchi and Soden 2007). Moreover, during that period both the ocean and atmosphere system were in an anomalous state. Scientists noted (Zhang and McPhaden, 2006) the tropical Pacific shallow ocean meridional overturning circulation (MOC) also weakened around that time. Hence to highlight the sensitivity to the choice, an arbitrary shift in baseline climatology from 1956–1997 to 1936–1975 is also performed (upper and lower panel).

The SLP gradient in ocean and land is likely to play a role in regulating ISM precipitation (Krishnamurthy and Kinter 2002) which comes into play via moisture transport. If separately the top and bottom panel are considered, it is clear that the contrast in SLP around sea and land shows similar patterns, irrespective of compositing for minimum or peak years. For Fig. 12.5a, b, there is a significant signal around

Darwin with an opposite significant region around the places of Indian land. Such climatological features have potential to modulate the intertropical convergence zone (ITCZ) and subsequently moisture convergence. If the focus is on the zero line around the Indian Ocean (bottom panel), the land-sea SLP contrast and influence on ITCZ are seen weaker in Fig. 12.5c, d; however, the gradient is the same. Also around Darwin of Australia, no significant signal is noticed. If the intensification of ITCZ was due to the Sun, the minimum year compositing would have been opposite (or at least different) to that from peak year compositing. Figure 12.5a, b clearly suggests that in general, there should be more monsoon precipitation during the period 1956–1997 (following the sign convention, as the anomaly is calculated about that period) if other factors are not considered into account, and, thus, the trend is playing an important role.

Another point to note here is that the region of significance for SLP shows an intensification during the compositing for peak year (Fig. 12.5, right panel) to that from respective compositing for minimum year (Fig. 12.5, left panel). As observed by Roy and Haigh (2010), such a feature might be linked with the preferential alignment of ENSO (cold phase) with solar peak years. It is evident from the difference between the right and left panel that it favours more intensification of ITCZ for solar max compositing and thus more rainfall. Such feature is same as the normal ISM and ENSO connection (during cold events of ENSO, there are more rainfall). Hence, not only the trend, the ENSO signal is also influencing ISM precipitation in compositing method that is attributed by van Loon and Meehl (2012) as entirely for solar. The overall analyses presented here indicate that while using the method of compositing to identify a solar signal in ISM, a caution should be taken.

Using peak year compositing solar signals was also detected in PDO and NAO by van Loon and Meehl (2014) which is also questionable. All such discussion suggests theories proposed based on solar peak year compositing to explain any solar signal are not based on robust evidence.

12.3   Difference in Winter Surface Climate Between Solar Minimum and Maximum

Figure 12.6 shows results from Ineson et al. (2011) for SLP data. Analysing reanalysis product from ERA-40/ERA-Interim and also model output they showed active solar years in 11-year cycle indicate positive NAO pattern.

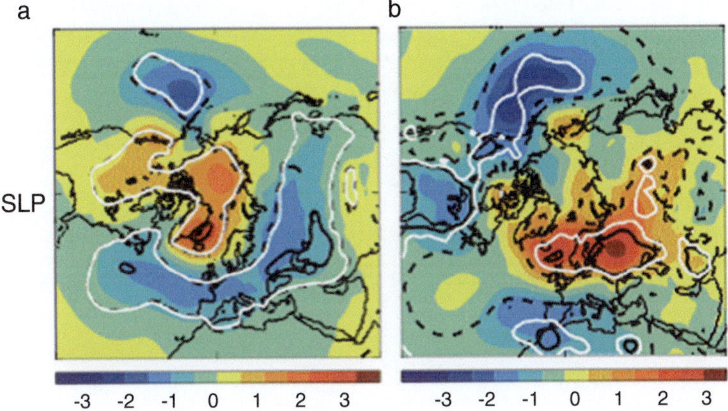

Fig. 12.6 Difference in sea level pressure (solar minimum-maximum) for the model (**a**) Model (left) and ERA-Interim/ERA-40 reanalysis (**b**) Reanalysis (right) (Ineson et al. 2011)

12.4   Sun (Using SSN) and NAO in Observation Using MLR Technique

12.4.1   Sun (Using SSN) and NAO in Two Different Time Periods (1856–1977) and (1878–1997)

Figure 12.7a shows the solar signature in SLP (spatial pattern) excluding other major factors, which are very likely to influence results. The right panel is for 1978–1997, whereas the left panel is for the earlier period of available data series. In both figures, a strong, positive signal around Aleutian Low (AL) is clearly detected, indicating the robustness of that signature noted in Roy and Haigh (2010). The right panel of Fig. 12.7 indicates a positive Arctic Oscillation (AO) pattern, whereas a clear dissimilarity is noticed around places of North Atlantic in the left panel. A positive NAO-like signature is clearly distinguished in the right panel, which is also noted in the study of Ineson et al. (2011). Surprisingly, that signature is opposite in nature in the left panel with a significant region mainly localised around the places of Greenland. Such observational analyses may raise doubt about any proposed mechanism, as discussed in Ineson et al. (2011).

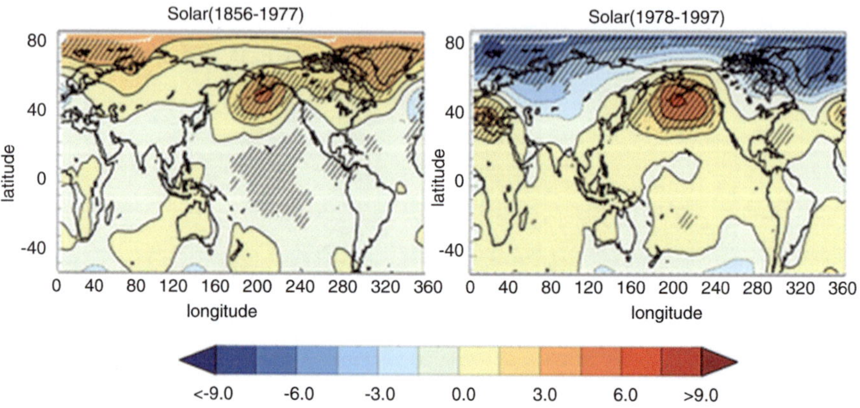

Fig. 12.7 Amplitudes of the components of sunspot number (SSN) variability for SLP (in hPa) during DJF using different independent factors as longer-term trend, OD and ENSO. The right panel is for the period (1978–1997) and the left panel for (1856–1977). Regions significant at the 95% level using a t-test are shaded, and negative contours are marked by dotted lines (Roy 2016)

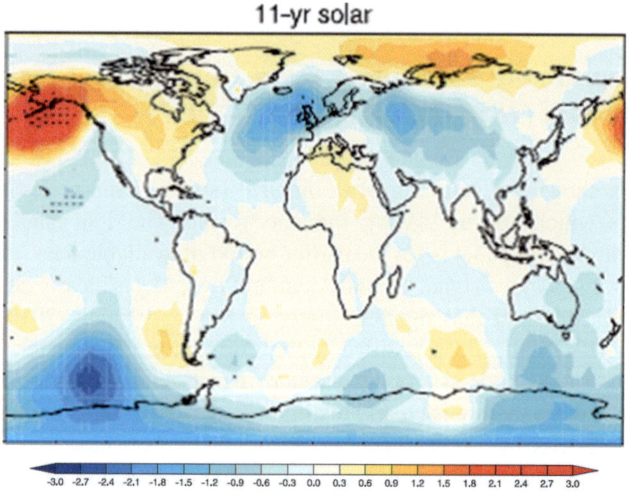

Fig. 12.8 The solar cycle signal (11 years) in HadSLP dataset (hPa) for 1870–2010 (DJF) using the regression technique. To have an estimate of solar minimum minus maximum, regression coefficients have been rescaled by the standard deviation and multiplied by the maximum trough-to-peak SSN. Black dots denote 95% significance using a student's t-test, while white dots for 99% level (Gray et al. 2013)

12.4.2   Sun (SSN) and NAO Longer Period (1870–2010)

Figure 12.8 shows result from Gray et al. (2013) using HadSLP2 data relating to the Sun and NAO connection for a longer period (1870–2010). It is again evident from the more extended period of record that NAO pattern is not significant, and it is even negative in signature. Roy et al. (2016) also could not detect significant solar (SSN) signal on SLP around the north Atlantic region that considered a period of 1900–2012. All those analyses raise doubts about any proposed mechanism on the Sun and NAO, as discussed in Ineson et al. (2011).

12.4.3   Sun (SSN) and NAO Lag Relationship

The recent studies by Scaife et al. (2013) and Gray et al. (2013) noted a solar lag relationship around places of the Azores High and discussed a mechanism based on their results. Using MLR technique Gray et al. (2013) examined the solar lag relationship and is shown in Fig. 12.9 (lag year 0 to lag year 3). Using HadSLP2 data for 1870–2010, it suggests a strong positive signature around Azores High in the 3-year lag from peak solar years.

But the lag relationship is also not robust and sensitive to the period chosen. The result of the overall time is mainly dominated by a period that has stronger signals.

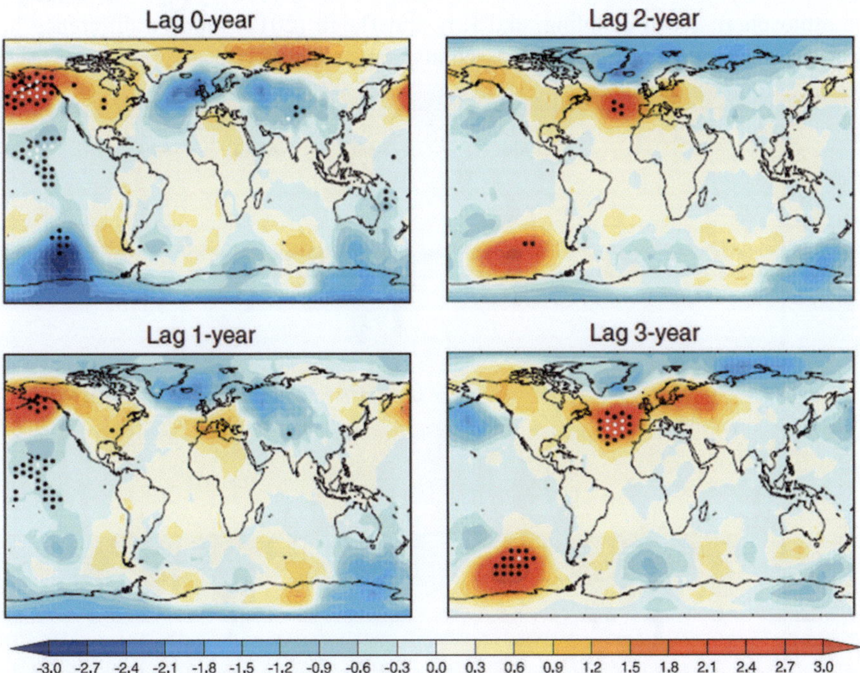

Fig. 12.9 The solar cycle signal (11 years) in HadSLP dataset (hPa) for 1870–2010 (DJF) using the regression technique. To have an estimate of solar minimum minus maximum, regression coefficients have been rescaled by the standard deviation and multiplied by the maximum trough-to-peak SSN. The mean SLP data lagging the SSN index by different numbers of years are shown. Black dots denote 95% significance using a Student's t-test, while white dots for 99% level (Gray et al. 2013)

If different periods say 1978–1997 and period before 1978 are used, the lag response around Azores High is not the same (Roy 2016). It suggests the lag signal may not be robust and needs to be explored further to address mechanism on Sun (using SSN) and NAO connection.

Hence, all the above discussions suggest that caution should be taken in proposing a mechanism based on any detected solar signal unless that signal is very robust.

12.5   AMO and PDO Relationship

Figure 12.10 shows different phases of the PDO and AMO in different time periods. It is observed that the PDO turned cold period around 1998 when global temperature hiatus started. The PDO switched to +ve around 1977 when an abrupt rise in global temperature started. Various studies indicated the year 1976/1977 as a climatic 'regime shift' (Miller et al. 1994; Meehl and Teng, 2014) because many physical conditions in the atmosphere and ocean including surface temperature changed abruptly during that period. It is also noticed in Fig. 12.10 that around 1960, AMO turned cold phase. Starting that period Vecchi and Soden (2007) observed a change in atmosphere-ocean coupling, and Roy and Haigh (2012) found a difference in Sun-ENSO connection and is discussed in details earlier.

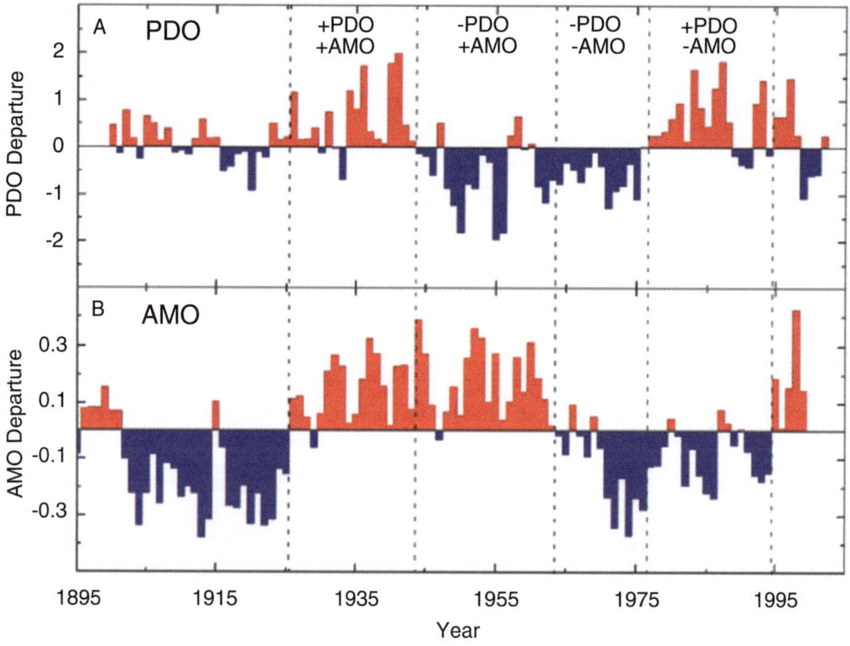

[McCabe, et al, 2004, Copyright (2004) National Academy of Sciences, U.S.A.]

Fig. 12.10 Combination of various AMO and PDO phases in different periods

Now it is known that the PDO and AMO control substantially the climate of the Earth, but the question remains unresolved as to what are the causes of AMO and PDO and can they be predicted. Many studies detected solar connection on both the AMO and PDO, but there are controversies and mechanism part is not clear.

In this chapter, we discussed the study of Meehl et al. (2009) relating to a solar influence on atmosphere-ocean coupling mentioning some limitations. Various issues related to detected solar signal using solar peak year compositing method are analysed. Solar 11-year cyclic variability and NAO connection suggested that strong positive solar signal as observed during the later period of the twentieth century is missing during the earlier period. Overall 150-year data could not indicate any significant connection between NAO and solar 11-year cyclic variability. The strong positive signature in SLP as noticed around Azores High in 2–3 year lag period is also sensitive to chosen time periods. It is important that all these are recognised so that theories presented for explaining any solar signature on climate are based on robust proof. Years of various phases of PDO and AMO are compared. Changes in phase of PDO around 1977 and 1998 are discussed concerning global climate.

References

Gray LJ, Scaife AA, Mitchell DM et al (2013) A lagged response to the 11 year solar cycle in observed winter Atlantic/European weather patterns. J Geophys Res Atmos 118:13405–13420. https://doi.org/10.1002/2013JD020062

Ineson S, Scaife AA, Knight JR et al (2011) Solar forcing of winter climate variability in the Northern Hemisphere. Nat Geosci 4:753–757. https://doi.org/10.1038/NGEO1282

IPCC (2007) Fourth assessment report of the intergovernmental panel on climate change. Cambridge University Press, Cambridge/New York

Krishnamurthy V, Kinter JL (2002) The Indian monsoon and its relation to global climate variability. In: Rodo X, Comin FA (eds) Global climate – current research and uncertainties in the climate system. Springer, Berlin, pp 186–236

Meehl GA, Teng H (2014) CMIP5 multi-model hindcasts for the mid-1970s shift and early 2000s hiatus and predictions for 2016–2035. Geophys Res Lett 41(5):1711–1716

Meehl GA, Arblaster JM, Branstator G, van Loon H (2008) A coupled air-sea response mechanism to solar forcing in the pacific region. J Climate 21(12):2883–2897

Meehl GA, Arblaster JM, Matthes K, Sassi F, van Loon H (2009) Amplifying the pacific climate system response to a small 11-Year solar cycle forcing. Science 325:1114–1118. https://doi.org/10.1126/science.117287

Miller J, Cayan DR, Barnett TP, Graham NE, Oberhuber JM (1994) The 1976–77 climate shift of the Pacific Ocean. Oceanography 7(1):21–26

Roy I (2014) The role of the sun in atmosphere-ocean coupling. Int J Climatol 34(3):655–677. https://doi.org/10.1002/joc.3713

Roy I (2016) The role of natural factors on major climate variability in Northern Winter. Preprints 2016, 2016080025 https://doi.org/10.20944/preprints201608.0025.v2

Roy I, Collins M (2015) On identifying the role of Sun and the El Niño Southern Oscillation on Indian Summer Monsoon Rainfall. Atmos Sci Lett 16:162–169

Roy I, Haigh JD (2010) Solar cycle signals in sea level pressure and sea surface temperature. Atmos Chem Phys 10(6):3147–3153

Roy I, Haigh JD (2012) Solar Cycle signals in the pacific and the issue of timings. J Atmos Sci 69(4):1446–1451. https://doi.org/10.1175/JAS-D-11-0277.1

Roy I, Asikainen T, Maliniemi V, Mursula K (2016) Comparing the influence of sunspot activity and geomagnetic activity on winter surface climate. J Atmos Sol Terr Phys 149:167–179

Scaife A, Ineson S, Knight JR, Gray L, Kodera K, Smith DM (2013) A mechanism for lagged North Atlantic climate response to solar variability. Geophys Res Lett 40(2):434–439

Tung KK, Zhou JS (2010) The pacific's response to surface heating in 130 yr of SST: La Niña–like or El Niño–like? J Atmos Sci 67:2649–2657

Van Loon H, Meehl GA (2011) The average influence of decadal solar forcing on the atmosphere in the South Pacific region. Geophys Res Lett 38(12) https://doi.org/10.1029/2011GL047794

Van Loon H, Meehl GA (2012) The Indian Summer Monsoon during peaks in the 11 year sunspot cycle. Geophys Res Lett 39:L13701

Van Loon H, Meehl GA (2014) Interactions between externally forced climate signals from sunspot peaks and the internally generated pacific decadal and North Atlantic Oscillations. Geophys Res Lett 41:161166. https://doi.org/10.1002/2013GL058670

van Loon H, Meehl GA, Shea DJ (2007) Coupled air-sea response to solar forcing in the pacific region during northern winter. J Geophys Res-Atmos 112:D02108. https://doi.org/10.1029/2006JD007378

Vecchi GA, Soden BJ (2007) Global warming and the weakening of the tropical circulation. J Climate 20:4316–4340

Zhang D, McPhaden MJ (2006) Decadal variability of the shallow pacific meridional overturning circulation: relation to tropical sea surface temperatures in observations and climate change models. Ocean Model 15(3-4):250–273

Chapter 13
The Arctic and Antarctic Sea Ice

Abstract This chapter covered Arctic and Antarctic climate. It discussed Arctic warming and focused on various related areas including Arctic Sea ice – last few years and recent decline. Recent behaviour of Antarctic Sea ice was also addressed.

Keywords Arctic sea ice · Antarctic sea ice · Sea ice extent · Sea ice trend

A strong warming trend in the Arctic during last one and half decades is noted as also seen from Fig. 13.1.

Melting of ice in the Arctic is shown in Fig. 13.2, with ice volume anomaly about 1979–2013 in 1000 km^3. It clearly suggests that there is a rapid decrease since 1977 (Fig. 13.3).

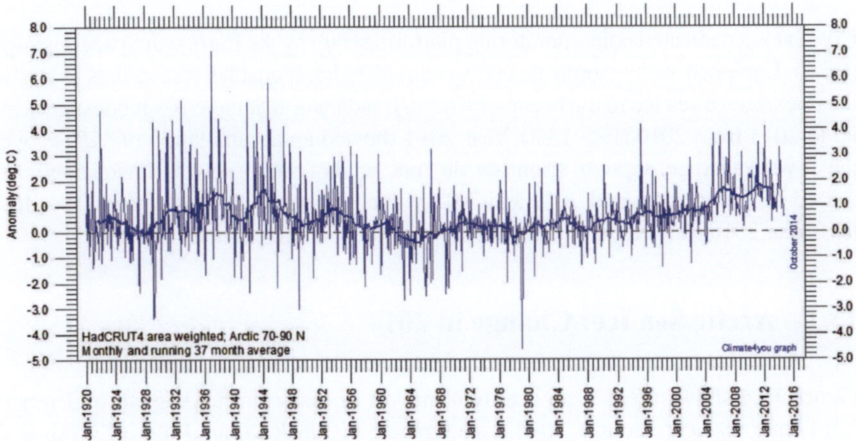

Fig. 13.1 Anomaly plot for HadCRUT4 area weighted Arctic 70–90° N monthly and running average anomaly. (Source: http://www.climate4you.com/Polartemperatures.htm#Arctic monthly surface air temperatures north of 70°N)

© Springer International Publishing AG, part of Springer Nature 2018

I. Roy, *Climate Variability and Sunspot Activity*, Springer Atmospheric Sciences, https://doi.org/10.1007/978-3-319-77107-6_13

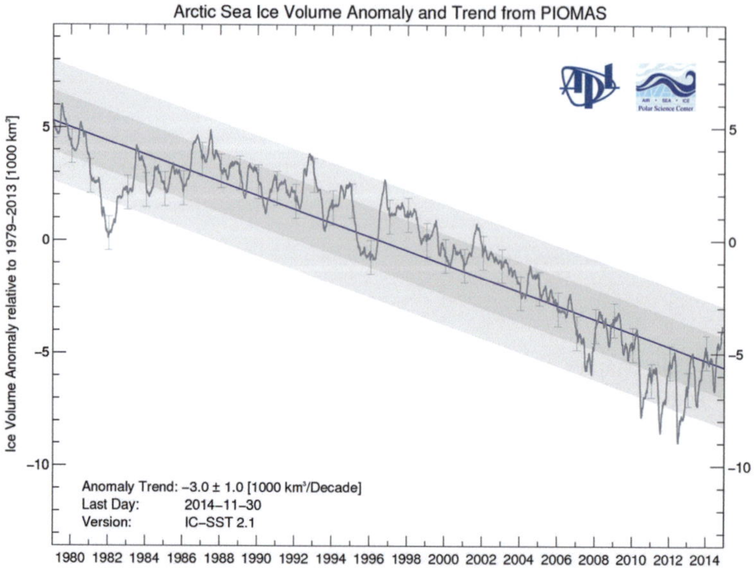

Fig. 13.2 Ice volume anomaly about 1979–2013 [1000 km³]. (Source: http://psc.apl.washington. edu/wordpress/wp-content/uploads/schweiger/ice_volume/BPIOMASIceVolumeAnomalyCurren tV2.1.png)

13.1   Arctic Sea Ice: Last Few Years

CryoSat is a dedicated polar monitoring platform set up by the European Space Agency (ESA). Equipped with a sophisticated system of radar, it enables researchers to study the thickness of sea ice in the ocean of Arctic. It indicates that there is a modest growth up to 2014 from 2010 (Fig. 13.3). Year 2014 showed an accumulation of 12% above the 5-year average, though scientists did not predict such growth. Interestingly, it started declining again and 2017 and 2018 winter experienced massive Arctic sea Ice loss. Year 2018 showed the maximum decline compared to previous winters.

13.2   Arctic Sea Ice: Change in 2014

Figure 13.4 shows there is an apparent growth in mass during August to October 2014 contrary to prediction. 2014 is shown here by black colour. Figure 13.5 shows Arctic Sea ice formation with thickness of ice and locations in a recent times (December 20th, 2013).

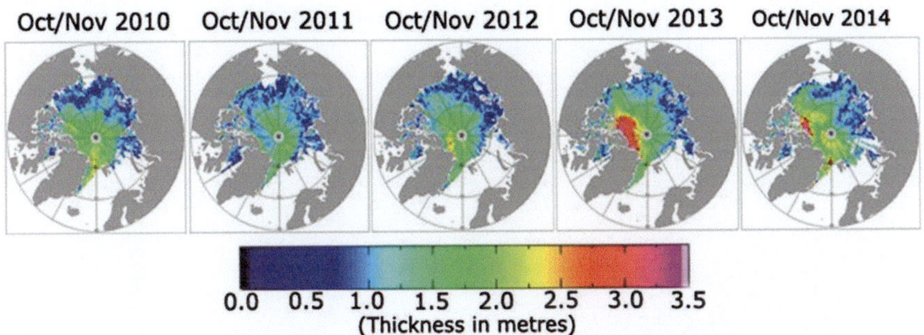

Fig. 13.3 Thickness of Arctic Sea ice (in metres) in the last few years during October–November. (Source: http://scitechdaily.com/images/Study-Shows-Volume-of-Arctic-Sea-Ice-Has-Increased.jpg)

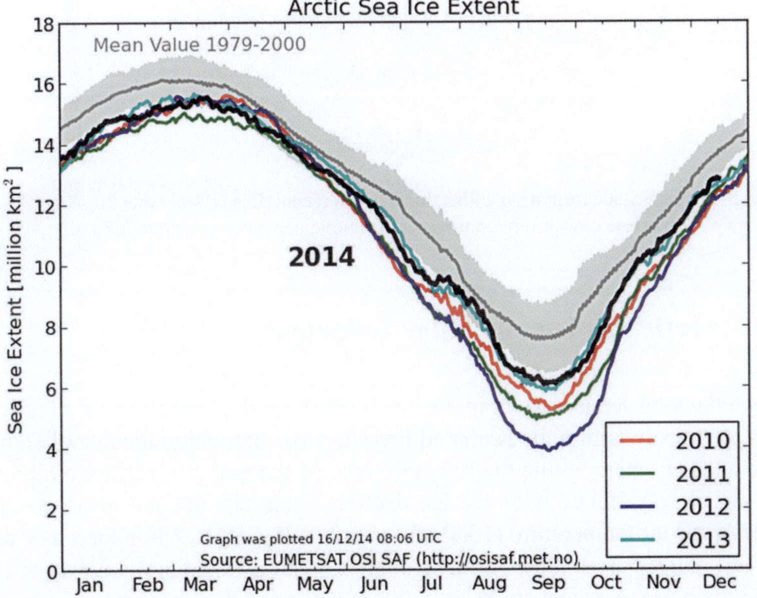

Fig. 13.4 Monthly value of sea ice extent is shown in various years (2010–2014) with different colours. Mean of 1979–2000 is shaded. (Source: EUMETSAT, OSI SAF (http://osisaf.met.no, graph plotted on 16/12/14.), also in http://kaltesonne.de/wp-content/uploads/2014/12/arktis2.gif)

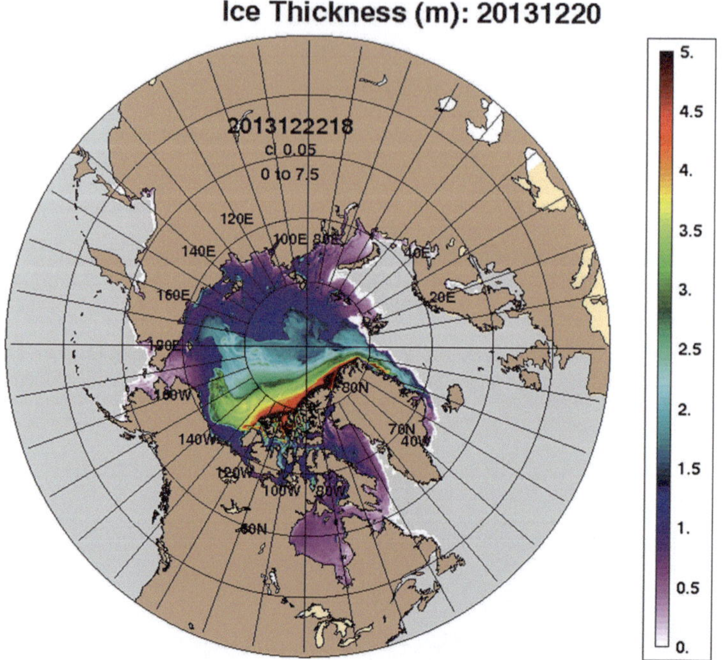

Fig. 13.5 Arctic Sea ice formation with a thickness in recent times (December 20, 2013). (Source: http://www7320.nrlssc.navy.mil/hycomARC/navo/arcticicen_nowcast_anim365d.gif

13.3   Arctic Sea Ice and Solar Influence

Though the total sea ice extent in the Arctic has been declining, the locations of ice losses are mostly within the Kara and Barents seas, and within the Sea of Okhotsk, and to a lesser extent within the Labrador Sea. As the trends in surface air temperatures are closely linked with sea ice decline, those features are also detected in winter Arctic air temperature (1000 mb) trends in Fig. 13.6. While there is a warming trend noticed around the Arctic, a cooling trend is noted in the Eurasian sector (shown by blue). A recent study (Roy 2018) discussed that those trends are related and current weaker solar cycle could be contributing to that.

It showed when the winter solar Sunspot Number (SSN) falls below average, there is a warming in the Arctic, though cooling in the Eurasian sector (Fig. 13.7). On the other hand, the situation reverses when SSN is above average. Such opposite features in the Arctic in different phases of SSN cycle are present from the lower troposphere to high up in the upper stratosphere and are linked with the Polar vortex and Arctic Oscillation phenomenon. That study discussed in spite of all other influences and complexities, it is still possible to segregate a strong influence from the sun. Heavy winter snowfall in Eurasia can cause colder than normal winter in all over Europe, like that happened in winter 2018.

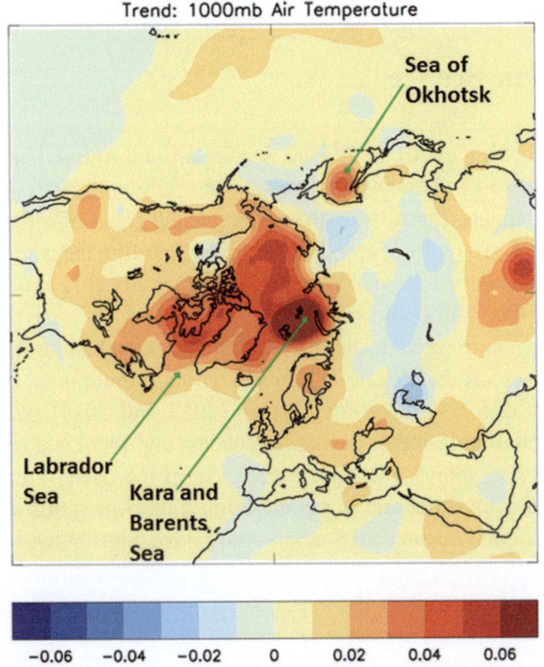

Fig. 13.6 The trend of winter air temperature (1000 mb) around Arctic (Roy 2018)

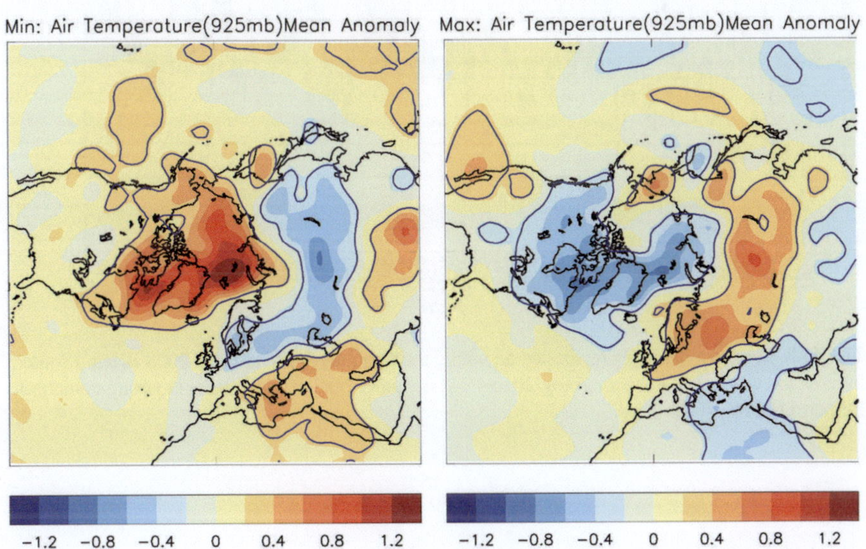

Fig. 13.7 The Composite anomaly of winter Arctic air Mean temperature (°C) at 925 mb (before detrending) for Solar Min (left) and Max (right) years. For Solar Min years, seasonal mean Sunspot numbers (SSN) are below average; while for Solar Max years, those are above average. After detrending, the sign of anomaly also remains similar. The base period is 1981-2010, and blue contour marks significant regions up to 95% levels of mean differences (Roy 2018)

13.4   Antarctic Sea Ice

In contrast to a melting of Arctic Sea ice, sea ice around Antarctica has expanded in recent years (Fig. 13.8). Reasons for this are not entirely understood.

Figure 13.9 indicates there is a strong decreasing trend of temperature in West Antarctica, and ice are melting at a dramatic rate. Satellite measurements by ESA's CryoSat-2 showed that each year West Antarctic Ice Sheet is losing enormous ice amounting more than 150 cubic kilometres.

In this chapter, there were discussions about Arctic and Antarctic snow cover. A sharp decline in ice around Arctic is noticed. Though there is an overall declining trend in the Arctic, a rise in recent 2 years (2013 and 2014) was not predicted. Antarctic is accumulating ice in past few decades, and there is a rising trend. West Antarctica, however, shows a dramatic decrease in ice cover. During 2017, there was, however, an overall decline in sea ice around the Antarctica and it is believed that the strong El Nino during 2015–2016 could have contributed to that.

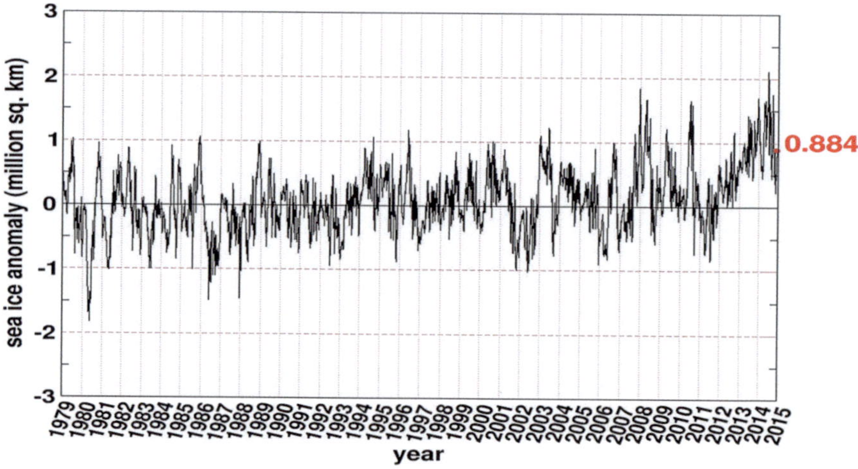

Fig. 13.8 Southern hemisphere sea ice anomaly, the anomaly from 1979 to 2008 mean. (Source: http://www.vencoreweather.com/blog/2016/4/11/215-pm-global-sea-ice-makes-a-strong-comeback, link on 11/4/16)

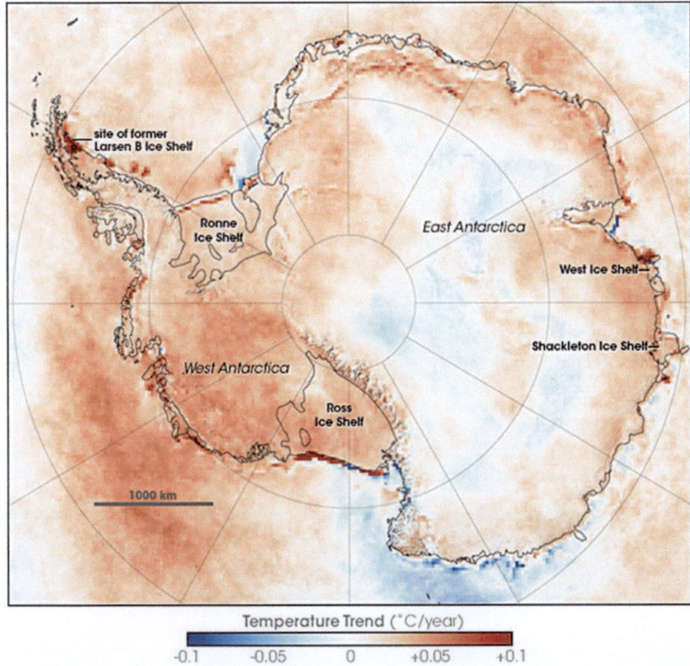

Fig. 13.9 Temperature trend in the Antarctic C/year. Warming is shown by red and cooling by blue. (Source: http://eoimages.gsfc.nasa.gov/images/imagerecords/8000/8239/antarctica_avhrr_81-07_lrg.pdf, link accessed on 2/4/18)

Reference

Roy I (2018) Solar cyclic variability can modulate winter Arctic climate. Sci Rep 8(1):4864. https://doi.org/10.1038/s41598-018-22854-0

Chapter 14
CMIP5 Project and Some Results

Abstract This chapter focused on Coupled Model Intercomparison Project, Phase 5 (CMIP5), and discussed various results. Starting from outlining very basic equations of global climate models (GCMs) as used in CMIP5 models, it described briefly about the aim and objective of the CMIP5 project. It is followed by discussion on various experiments, historical and RCP (Representative Concentration Pathway) situation. The global temperature generated using CMIP5 models was compared with observation. Indian Summer Monsoon and El Niño Southern Oscillation in CMIP5 models were analysed in the historical and RCP situation, and few areas of agreement and disagreement were discussed. Few results from the atmospheric version of CMIP5 models (AMIP5) and Phase 3 of CMIP5 experiments (CMIP3) were also presented. Some stratospheric features are shown well captured by models.

Keywords CMIP5 · AMIP5 · Historical Scenario · RCP Scenario · RCP 2.6 · RCP 4.5 · RCP 8.5 · High Top Model · Low Top Model

14.1   Global Climate Models (GCMs): Basic Equations

Global climate models (GCMs) are simulations generated from computer of the state of the atmosphere-ocean system. Using numerical representation of processes, GCMs study climate system and their complex interactions alongside its sensitivity and evolution to external or internal forcing. As the atmosphere is a compressible, continuous fluid in a rotating planet, using basic laws of physics, it is possible to develop an understanding of the global atmospheric circulation and their main features. It involves various complex equations, and to discuss those extensively is beyond the scope of this book. Here is a very brief and general overview. These are the main basic equations used:

Equation of State

$$PM = \rho RT$$

© Springer International Publishing AG, part of Springer Nature 2018

I. Roy, *Climate Variability and Sunspot Activity*, Springer Atmospheric Sciences, https://doi.org/10.1007/978-3-319-77107-6_14

M is molecular weight, *P* is atmospheric pressure, ρ is density, *T* is temperature and *R* is universal gas constant.

Navier-Strokes Equation

In rotating frame with angular velocity Ω after scale analysis reduces to

$$fk \times U_h = -\frac{1}{\rho} \nabla_h P$$

h represents the horizontal direction, which is the dominant. *U* is vector velocity, *k* is the unit vector in the vertical path and Coriolis parameter is *f*, where
$f = 2\Omega \sin \Phi$, Φ is latitude. Other factors are defined earlier.

Continuity Equation

$$\frac{\partial \rho}{\partial t} = -\nabla \cdot (\rho u)$$

u is local velocity. Left-hand side represents the rate of variation in density with time.

Thermodynamic Equation

$$\frac{DT}{Dt} = Q + \frac{1}{\rho Cp} \frac{DP}{Dt} \qquad \text{Where,} \quad \frac{D}{Dt} = \frac{\partial}{\partial t} + u.\nabla$$

Left-hand side represents rate of change following air parcels, which is usually known as Lagrangian method. *Q* is the heating rate associated with diabatic processes, i.e. tropospheric latent cooling/heating and stratospheric radiative processes etc. *Cp* is specific heat at constant pressure.

14.2   CMIP5 Project

A latest joint initiative around the world among various modelling groups of GCMs coordinated similar experiments that comprises the 5th phase of Coupled Model Intercomparison Project (CMIP5) [Link: http://cmip-pcmdi.llnl.gov/cmip5/experiment_design.html]. It is an improved version to that from earlier version, CMIP3. The details are described in Taylor et al. (2012). The CMIP5 models are capable of simulating ENSO-like variability in the equatorial Pacific and include an interactive ocean. They are capable of simulating interannual SST variability with realistic amplitude in the eastern and central equatorial Pacific as discussed by Bellenger et al. (2014). They also showed more CMIP5 models compared to CMIP3 group of models suggest a practical range of ENSO frequencies in the band of 2–7-year around east equatorial Pacific. Approximately half of the CMIP5 models show SST anomalies that peak during November to January, as seen in observation (Bellenger et al. 2014).

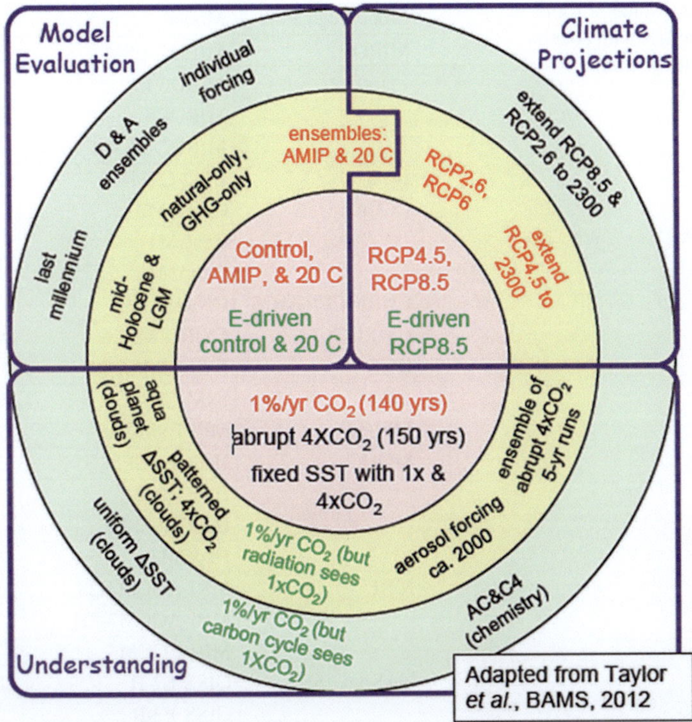

Fig. 14.1 Various CMIP5 experiments for historical and RCP scenario

Few Aims of CMIP5 Project
1. Assessing the mechanisms responsible for model differences
2. Poorly understood feedbacks associated with clouds, etc.
3. Exploring the capability of models to predict climate on decadal scales
4. Determining why similarly forced models to produce a range of responses

14.3 Experiments: Historical and RCP (Representative Concentration Pathway) Scenarios

Historical experiments are run for 155 years from 1850 to 2005 with all forcing. RCP85 means Representative Concentration Pathway (RCP) that leads to an approximate radiative forcing of 8.5 W/m². Future Projection for 2006–2300 are run forced by RCP8.5 scenario. There are other RCP scenarios as shown in Fig. 14.1 as RCP2.6, RCP4.5, etc. Different CMIP5 experiments are run for specific purposes such as model evaluation; climate projections and understanding of processes are shown in Fig. 14.1.

Table 14.1 Name of some CMIP5 models with country and modelling centre

Modelling centre	Model	Country
MIROC	MIROC-ESM	Japan
	MIROC4h	
	MIROC5	
	MIROC-ESM-CHEM	
BCC	BCC-CSM1.1	China
CCCma	CanESM2	Canada
	CanCM4	
	CanAM4	
CNRMCERFACS	CNRM-CM5	France
CSIROQCCCE	CSIRO-Mk3.6	Australia
IPSL	IPSL-CM5A-LR	France
INM	INM-CM4	Russia
MPI-M	MPI-ESM-LR	Germany
MOHC	HadGEM2ES	UK
	HadGEM2A	
	HadGEM2-CC	
	HadCM3	
MRI	MRI-AGCM3-2S	Japan
	MRI-AGCM3-2H	
	MRI-CGCM3	
NASA-GISS	GISS-E2-H	USA
	GISS-E2-R	
NCAR	CCSM4	USA
NCC	NorESM1-M	Norway

14.4 Some CMIP5 Models

Table 14.1 shows the name of some CMIP5 models with country and modelling centre. For example, in UK Met Office Hadley Centre (MOHC) runs four CMIP5 models which are HadGEM2ES, HadGEM2A, HadGEM2-CC and HadCM3.

14.5 Temperature in CMIP5 and Observation

Observations are shown by circles (average 4 balloon datasets) and squares (average 2 satellite datasets)

Figure 14.2 shows the temperature of tropical mid-troposphere for 73 CMIP5 models. The average of all models is marked by red and circles, and squares indicate various observations. Circles are for an average of 4 balloon data, whereas average of 2 satellite datasets are shown squares. From Fig. 14.2, it is quite clear that the deviation between observation and ensemble mean of model results increased since

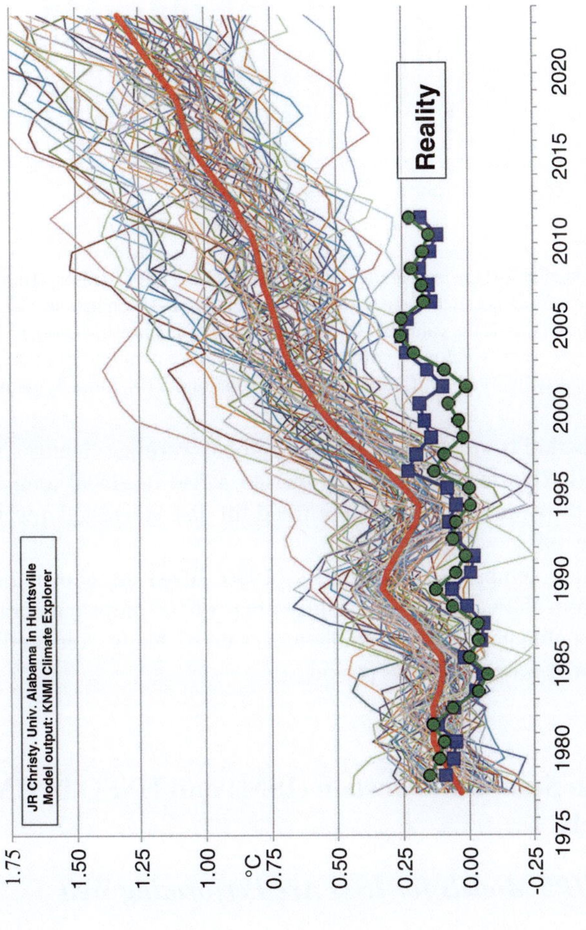

Fig. 14.2 The temperature of tropical mid-troposphere 20S–20N for 73 CMIP5 models. (Source: https://ktwop.files.wordpress.com/2013/10/73-climate-models_reality.gif, uploaded on 30/12/2017. Credit: JR Christy, University of Alabama, Huntsville-model output from KNMI)

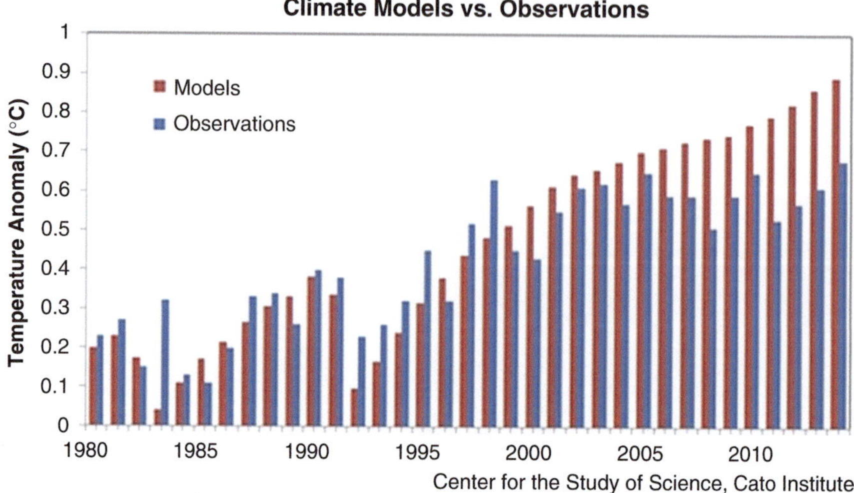

Fig. 14.3 Global annual surface temperature anomalies (1980 to 2014). Observations from NOAA (blue) are anomalies from twentieth-century average. Average of 108 climate models are shown in red. For NOAA observations, the value of 2014 is an average of January to October. (Source: http://www.cato.org/blog/current-wisdom-record-global-temperature-conflicting-reports-contrasting-implications. uploaded on 30/12/17, Cedit: © The Cato Institute 2014. Used by permission)

1998 and that period from 1998 is known as global warming 'hiatus'. There was a recent strong El Nino in 2015–'16 that caused a rise in global temperature. But inspite of that, the observed temperature trend till date suggested a warming trend slowdown since 1998.

Annual surface temperature anomalies (1980–2014) are shown in Fig. 14.3. Observations from NOAA are in blue, and average of 108 climate models are in red. It is evident that climate model ensemble means consistently over-predict observation, during global warming hiatus period.

14.6 Indian Summer Monsoon (ISM) and ENSO in CMIP5 Models

14.6.1 CMIP5 Models for ISM Are Performing Well

A recent study (Roy and Tedeschi 2016) analysed CMIP5 model outputs to improve the understanding of ISM rainfall. Using CMIP5 and CMIP3 models, Jourdain et al. (2013) discussed ISM rainfall and its connection with ENSO in detail. Figure 14.4 shows climatology of ISM is captured well in some CMIP5 models.

The boreal summer (June to September, JJAS) climatological precipitation has a widespread maximum over central-northern India that includes Indo-Gangetic Plain. Considering a box in that region (76° to 87°E, 20° to 28°N), Bollasina et al. (2011) detected a decreasing trend for ISM precipitation, during the period 1940–

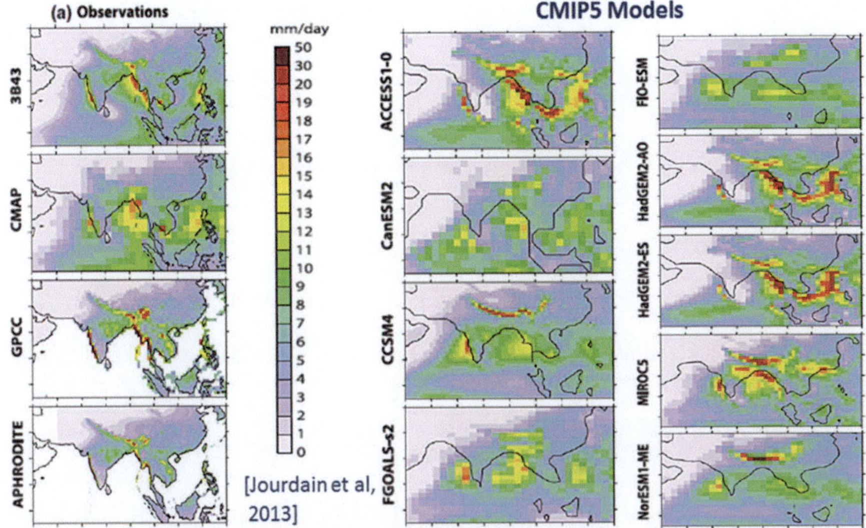

Fig. 14.4 Rainfall climatology in observations (left) and in historical CMIP5 simulations (right) (Jourdain et al. 2013)

2005, using NOAA GFDL CM3 model for all forcing (natural and anthropogenic) condition. The Climate Research Unit (CRU) data also showed a marked reduction from the 1950s to the end of the twentieth century (significant at the 95% level), which is compared with their model results by Bollasina et al. (2011) and shown in Fig. 14.5. Here their model results agree with the observed ISM behaviour.

14.6.2   CMIP5 Models for ISM Not Performing Well

Following the work of Bollasina et al. (2011), here the focus is to investigate the spatial and temporal behaviour of the ISM precipitation in the same box region (76° to 87°E, 20° to 28°N). It considers a similar period (1940–2005) and uses various CMIP5 model outputs. Figure 14.6 focuses on the spatial pattern while Fig. 14.7 on the temporal variation. Those figures clearly suggest very widespread behaviour among models.

14.6.2.1   Spatial Patterns in the Box Region

To investigate the nature of variation in spatial pattern, mean values of precipitation during the period 1940–2005 is subtracted to that from the average for the period 1986–2005. Such anomaly calculation can indicate whether there is a rising or decreasing trend during a later period. For example, if the calculated anomaly indicates positive, that suggests there is a decrease in rainfall during the later period

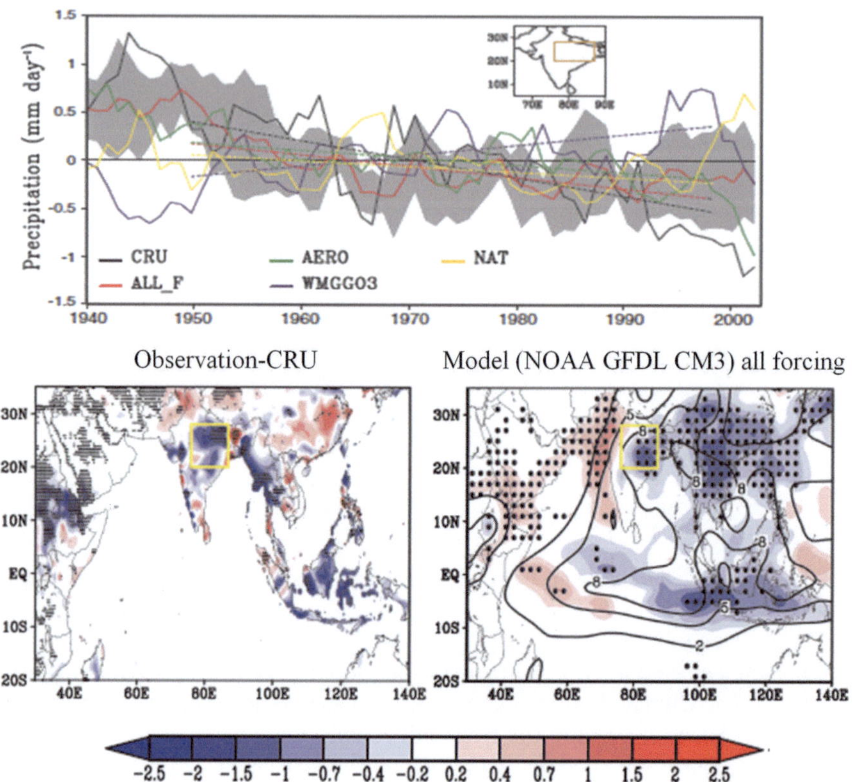

Fig. 14.5 The temporal pattern for 5-year running mean JJAS rainfall (relative 1940–2005) over central-northern India (box region marked by yellow) is shown on top. Anomalies are calculated as deviations from the 1950 to 2005 climatology. Observation from CRU is shown by black, while results using model NOAA GFDL CM3 for all forcing is shown by red. The spatial pattern of 1950–1999 least-square linear trends of JJAS average precipitation is shown in the bottom panel. The right plot shows the ensemble mean for all forcing and the left for observation from CRU. The black dots mark the grid point where trend exceeds the 95% significant level. (From Bollasina et al. 2011. Reprinted with permission from AAAS)

(1986–2005) as anomaly calculated about that period. Figure 14.6 shows precipitation anomaly using various CMIP5 models. A dark black rectangle marks the box region of Bollasina et al. (2011) in central northeast India. The red colour in that area suggests a rise in the precipitation during the later period (as an anomaly about that period). In other words, to compare with Bollasina et al. (2011), it can be said there is a rising trend. Unlike their observation, it is apparent from Fig. 14.6 that there is no clear consensus among models. It shows models dominated by rising trend in row 1 to models dominated by decreasing trend in the last row, with models without any indications of a trend in between.

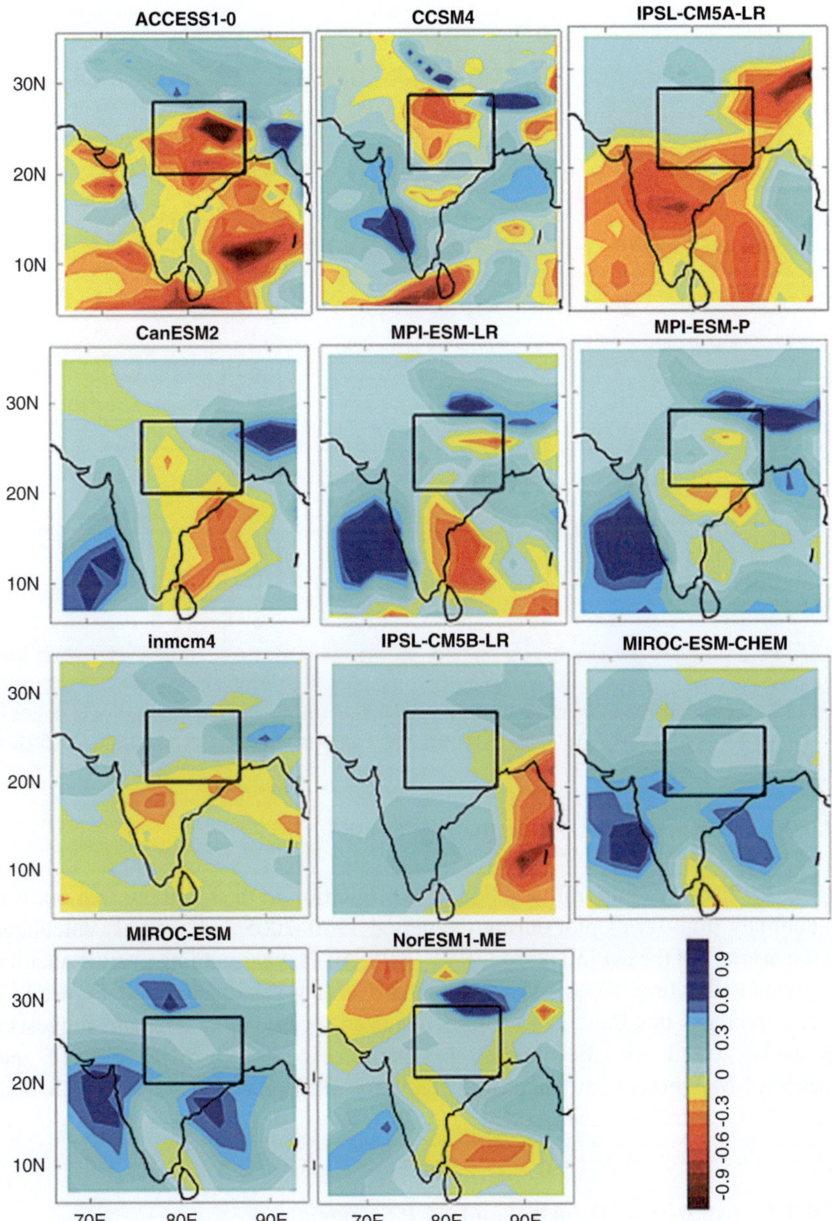

Fig. 14.6 Spatial pattern of ISM precipitation (1940–2005) anomaly (mm/day) about the period 1985–2005 (Roy 2017)

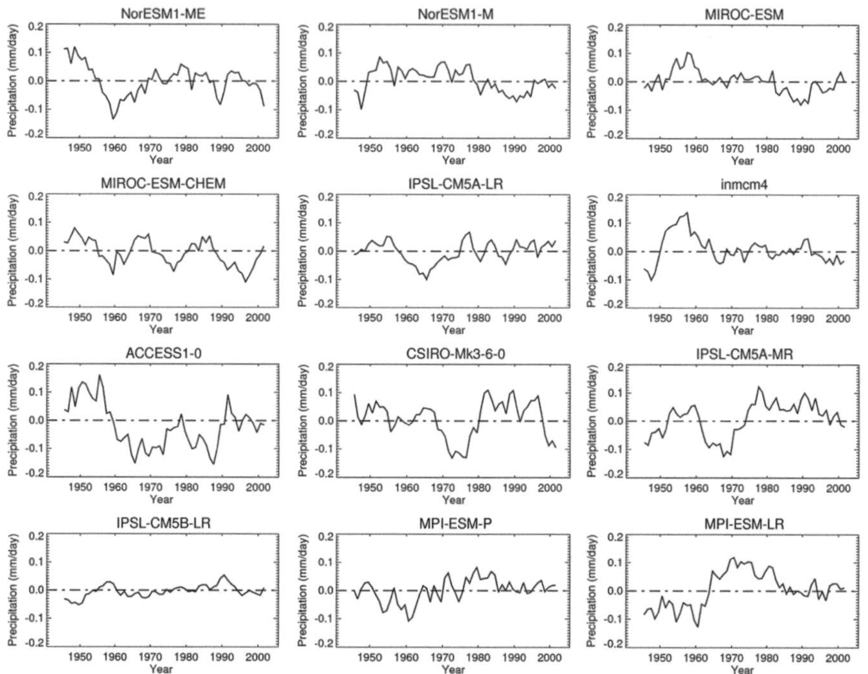

Fig. 14.7 Eleven-year running mean of ISM precipitation (historical run) time series is shown in box region for last 50 years. Anomaly is shown about an average of the period 1940–2005 (Roy 2017)

14.6.2.2   Temporal Patterns in the Box Region

Figure 14.7 shows temporal variation of precipitation in the same box region. It is an anomaly time-series plot during the period 1940–2005 and anomaly calculated on the average of the overall period 1940–2005. An 11-year running average method is applied to that time series, to have a clearer overview relating to trend. Figure 14.7 clearly indicates that there is no common consensus among models and suggests a decreasing trend for the models in mcm4, Access1–0, MIROC-ESM and NorESM1-M whereas an increasing trend for IPSL-CM5B-LR, MPI-ESM-LR, MPI-ESM-P and IPSL-CM5A-MR.

14.6.2.3   Combined Historical and RCP Scenario

To have a clearer overview relating to a future scenario in the box region, precipitation in the RCP8.5 situation is also plotted in Fig. 14.8. Here, overall historical period (1850–2005) is also considered. The anomaly period is considered as the period 1985–2005. In this plot, the historical period is shown by black, whereas RCP8.5 scenario (period 2005–2100) is shown in blue. Eleven-year smoothing is

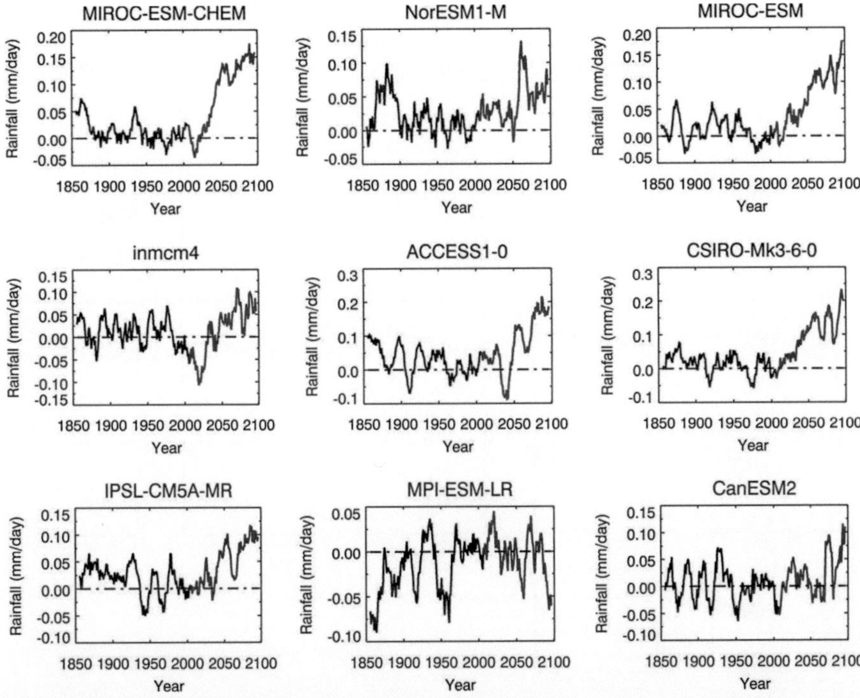

Fig. 14.8 Eleven-year running mean of ISM precipitation time series is shown in the box region for historical (black) and RCP (blue) scenario. Anomaly is calculated about the period 1985–2005 (Roy 2017)

applied to the whole time series. It is clear that though there is a rising trend in precipitation in the box region among most of the models in a future scenario, still there are models, e.g. MPI-ESM-LR, which show a decreasing trend.

14.6.2.4   Niño 3.4 in RCP Scenario

Figure 14.9 shows time series of SST in Niño 3.4 region in the RCP8.5 scenario. Anomaly is calculated about the period 1985–2005. Y-axis range is kept constant to show the difference in variability and trend among models.

It is apparent from Fig. 14.9 that for some models, Niño 3.4 temperature is highly variable than others. For example, model MIROC5 indicates high variable Niño temperature. Other models, those also suggest similarly, are GFDL- ESM2M and FGOALS-s2. Some models in that figure can be identified showing least variable Niño temperature, e.g. inmcm4, MIROC-ESM and MIROC-ESM-CHEM. The variability suggests similarly during the historical period of respective models (hence not shown). Years of peak and trough of Niño variability differ among models during historical as well as RCP scenario.

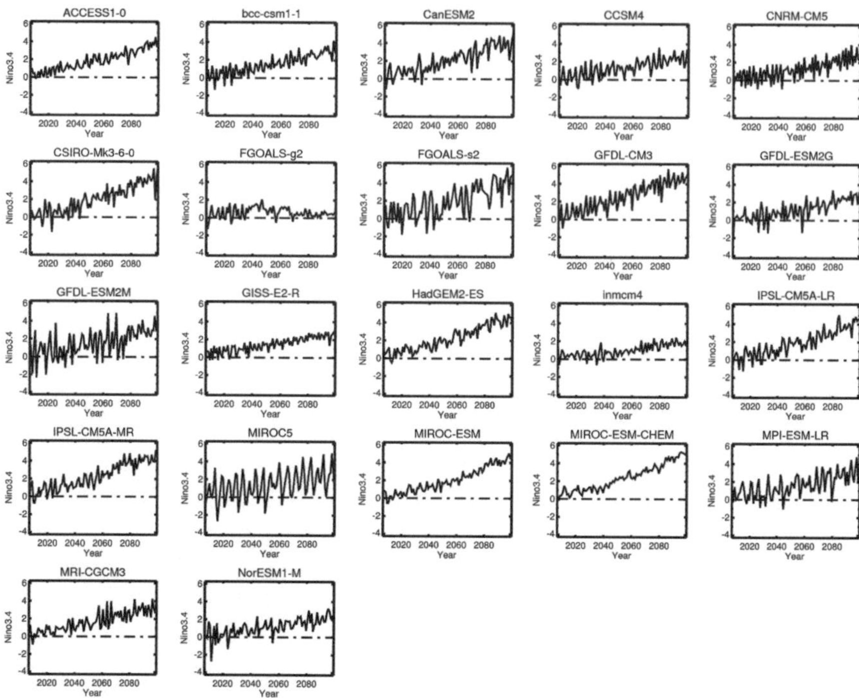

Fig. 14.9 Niño 3.4 time series for various models during JJAS for RCP scenario anomaly calculated about the period 1985–2005 (Roy 2017)

Regarding trend in Niño temperature for RCP scenario, model results also vary (Fig. 14.9); some models show the least trend, e.g. inmcm4, NorESM1-M, while some suggest a larger trend, e.g. GFDL-CM3, MIROC-ESM and MIROC-ESM-CHEM. In spite of the varied degree of a trend (least trend or larger trend), all models show a rising trend in RCP scenario with one exception for the model FGOALS-g2. This particular model fails to indicate anything about Niño trend and thus needs to be investigated further for such discrepancy. Such issues are dealt by respective 'Model Evaluation Group'.

Table 14.2 is formulated to show the correlation between ISM rainfall and Niño 3.4 temperature. It calculates the value of correlation during historical as well as RCP scenario. For precipitation, the region of the whole Indian subcontinent is considered separately alongside the same box area. The trend is removed from both the time series before doing calculations. Negative signature in the correlation value is clearly noticed for all cases irrespective of historical or RCP scenario. Thus CMIP5

Table 14.2 ISM and ENSO correlation using various CMIP5 models (in the whole Indian subcontinent and the box region) during historical and RCP scenario, after removing trend

| Model | Correlation: Rainfall vs. Niño3.4 | | | |
| | Whole India | | Box | |
	(Historical)	(RCP)	(Historical)	(RCP)
NorESM1-M	−.78	−0.71	−.47	−.40
MIROC-ESM	−.23	−0.23	−.03	−.10
Inmcm4	−.44	−0.39	−.37	−.27
ACCESS1–0	−.29	−0.16	−.11	−.15
CSIRO-Mk3–6-0	−.26	−0.30	−.24	−.39
IPSL-CM5A-Mr	−.66	−0.54	−.60	−.50
MPI-ESM-LR	−.35	−0.50	−.33	−.52
CanESM2	−.49	−0.65	−.26	−.31

models perform well to capture the usual ISM rainfall and Niño temperature behaviour, suggesting less rainfall during El Niño phase and more during La Niña (Turner et al. 2005; Maity and Kumar 2006). Results from few models are presented in Table 14.2, though the rest of the models also suggest similarly.

14.6.2.5 Precipitation in RCP Scenario During JJAS

Models are considered as the primary viable tool to predict future scenario. Here some situations for future are analysed for precipitation and presented in Fig. 14.10. Precipitation time series in the land region during JJAS in RCP 8.5 scenario is calculated for the period 2005–2100 about the average of 1985–2005. The top panel shows that of the whole globe, land region, whereas the bottom one for the Indian subcontinent.

For the world (Fig. 14.10, top panel), there is a clear rising trend noticed. The spread among models increases with time showing maximum range at around the year 2100. However, in the case of the Indian subcontinent (bottom panel), it is not possible to identify any clear trend pattern among model results. Moreover, throughout the period, there lies a large spread. It indicates CMIP5 models are unable to suggest precisely about temporal evolution of ISM rainfall, and models vary to a greater degree among each other.

14.6.3 Models: CMIP5, AMIP5 and High Top, Low Top

Inside CMIP5, there are simulations those use atmospheric global circulation models' (AGCM) outputs. Those are known as Atmospheric Model Intercomparison Project (AMIP5), which have prescribed SST (Hurrel et al. 2008). CMIP5 simulations are available for the period 1861–2005 while AMIP5 for 1979–2008. Models having

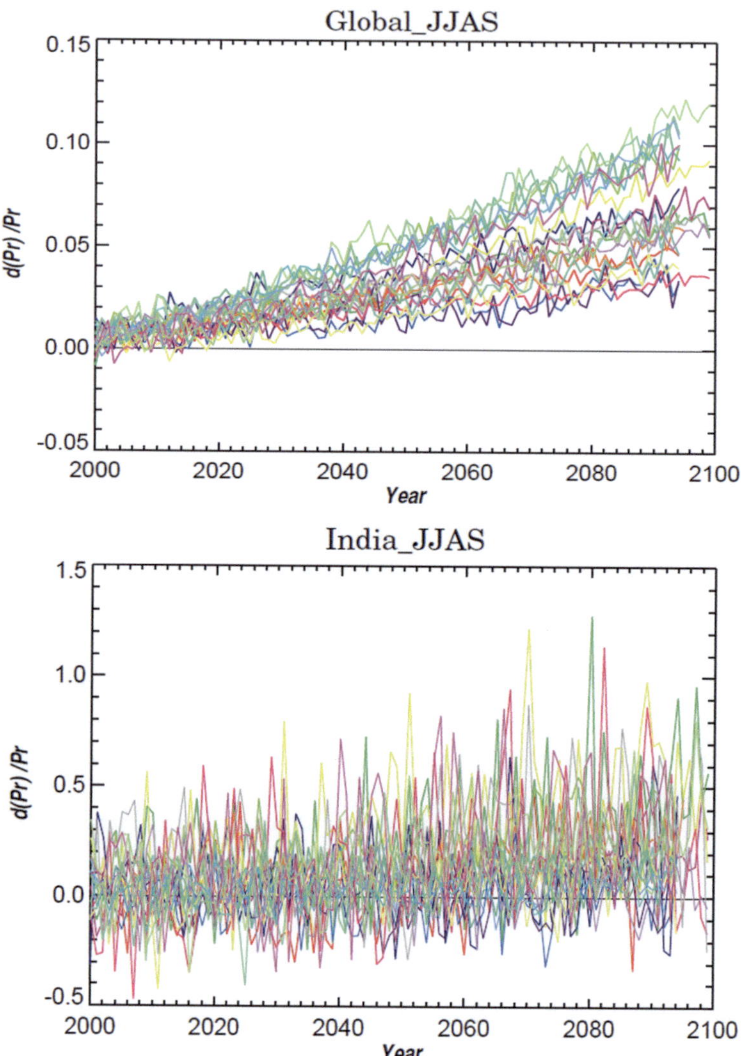

Fig. 14.10 Precipitation (Pr) time series during JJAS using various CMIP5 models for all over the globe (top) and for all over Indian subcontinent (bottom). Time series plotted about the period 1985–2005 (Roy 2017)

both the versions of AMIP5 and CMIP5 are shown in Table 14.3 (altogether 23 models are shown here). Models are also sometimes segregated as low or high top as demonstrated by L or H in Table 14.3 (last column). High-top models have few model layers in the stratosphere and have upper lids up to the stratopause (1 hPa). Eight of those models are considered as 'high top' while 15 are 'low top' in Table 14.3. Several studies suggested that such separation better captures extratropical features and related teleconnections (Hurtitz et al. 2014; Charlton-Perez et al. 2013).

Table 14.3 Various modelling centres, with the name of model types, CMIP5 and AMIP5, separated as low or high top

Model centre	Model type		High top (H)/
	CMIP5	AMIP5	low top (L)
CSIRO-BOM, Australia	ACCESS1.0	ACCESS1.0	L
	ACCESS1.3	ACCESS1.3	L
BCC, China	BCC-CSM1.1	BCC-CSM1.1	L
	BCC-CSM1.1(m)	BCC-CSM1.1(m)	L
GCESS, China	BNU-ESM	BNU-ESM	L
CCCMA, Canada	CanESM2	CanAM4	L
NCAR, USA	CCSM4	CCSM4	L
CMCC, Italy	CMCC-CM	CMCC-CM	L
CNRM-CERFACS, France	CNRM-CM5	CNRM-CM5	L
CSIRO-QCCCE, Australia	CSIRO-Mk3.6.0	CSIRO-Mk3.6.0	L
LASG-CESS, China	FGOALS-g2	FGOALS-g2	L
LASG-IAP, China	FGOALS-s2	FGOALS-s2	L
INM, Russia	INM-CM4	INM-CM4	L
MIROC, Japan	MIROC5	MIROC5	L
NCC, Norway	NorESM1-M	NorESM1-M	L
NOAA-GFDL, USA	GFDL-CM3	GFDL-CM3	H
MOHC, England	HadGEM2-CC	HadGEM2-A	H
NASA-GISS, USA	GISS-E2-R	GISS-E2-R	H
IPSL, France	IPSL-CM5A-LR	IPSL-CM5A-LR	H
	IPSL-CM5A-MR	IPSL-CM5A-MR	H
MPI-M, Germany	MPI-ESM-LR	MPI-ESM-LR	H
	MPI-ESM-MR	MPI-ESM-MR	H
MRI, Japan	MRI-CGM3	MRI-CGCM3	H

14.6.4   Precipitation Composites- El Niño: (CMIP5 vs. AMIP5)

Figure 14.11 shows precipitation composites for El Niño (EN). The comparison is made for one typical low-top (CCSM4) and high-top CMIP5 model (HadGEM2-CC) with the observation of ISM rainfall from Global Precipitation Climatology Project (GPCP). The composites in ENSO Modoki (ENM) years are in the right panel; left is for ENSO Canonic (ENC) and the middle one for ENSO Canonic and Modoki (ENCM) years. These two models are chosen as they well reproduce climatology of ISM precipitation (Jourdain et al. 2013). Results from AMIP5 low-top model CCSM4 are also presented in the bottom panel of Fig. 14.11. As seen, the model results for compositing related to ENSO-ISM teleconnection deviate a lot from observation.

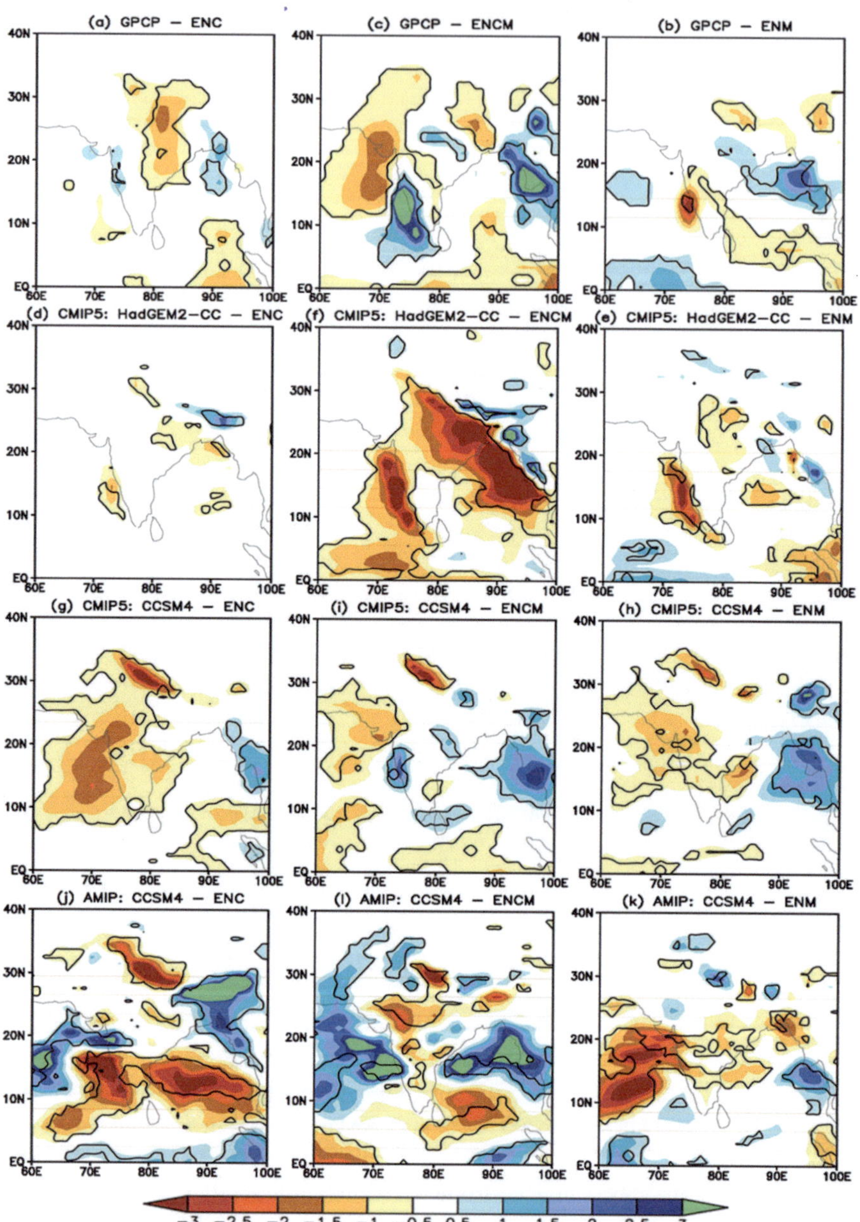

Fig. 14.11 Precipitation composites for EN, CMIP5 vs. AMIP5, comparing one typical low-top (CCSM4) and high-top CMIP5 model (HadGEM2-CC) with observation (GPCP). The top panel shows observation and the bottom AMIP low-top CCSM4 model. The right is for composites in ENSO Modoki (ENM), left for ENSO Canonic (ENC) and the middle one for ENSO Canonic and Modoki (ENCM) years (Roy et al. 2017)

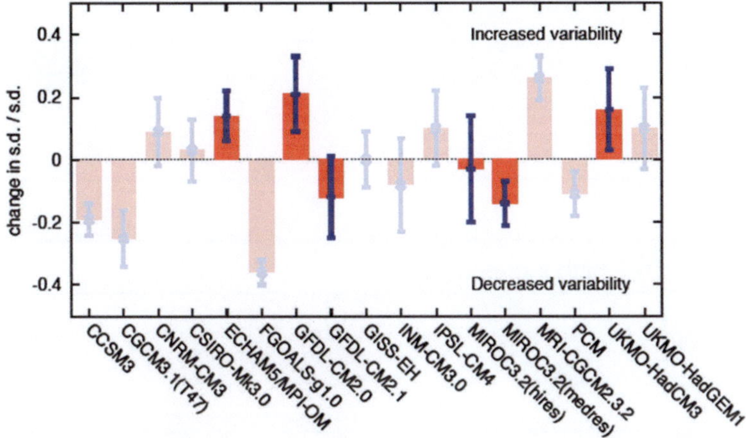

Fig. 14.12 Change in future ENSO variability in CMIP3 models (Collins et al. 2010)

14.6.5   Changes in ENSO Variability 2050–2100 in CMIP3 Experiments

Figure 14.12 shows how various CMIP3 models perform regarding future (2050–2100) variability. There is no consensus among models – some show increased variability, while some indicate decreased variability.

14.6.6   Stratospheric Features in CMIP5: Low and High Top Models

Using ECMWF reanalyses and CMIP5 models, the downward propagation of negative NAM events is shown in Fig. 14.13. Such downward propagation follows Baldwin and Dunkerton (2001). Tropospheric responses in low-top models are found to weak as seen in Fig. 14.13

14.6.7   Simulated and Observed Stratospheric Temperature

In Fig. 14.14, global mean (82.5°N–82.5°S) annual lower stratospheric temperature (TLS) anomalies from 1978 to 2011 mean are shown (top). Black colour shows dataset from remote sensing solutions; blue is the mean of low-top models, while red is for high top. Pink and blue shades show ~5–95% ensemble range. In the bottom panel, corresponding trends of mean zonal temperature over the same time is shown with volcano years excluded.

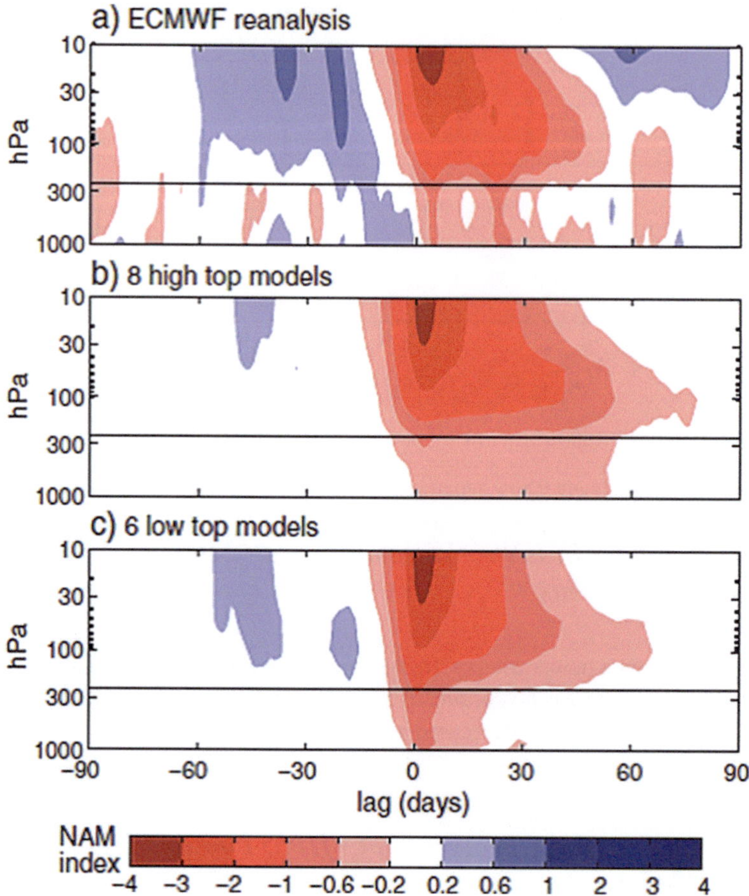

Fig. 14.13 In ECMWF reanalyses and CMIP5 models, downward propagation of negative NAM events is shown. The multimodel composites are based on all events in the CMIP5 historical simulations in either the low- or high-top ensembles but weighted so that every model makes an equal share. Events are defined by instances when the NAM index at 10 hPa drops below −3 standard deviations, and the composites are constructed from NAM anomalies in the 90 days proceeding (positive lags) and preceding (negative lags) the dip at 10 hPa (Charlton-Perez et al. 2013)

Figure 14.14 (top) showed the good agreement between observed and simulated lower stratosphere in both low- and high-top simulations. The positive temperature anomaly is overestimated in both sets of simulations, following Pinatubo eruption 1992. They suggested that due to this model's stratospheric aerosol scheme, it overestimated the stratospheric temperature response to Pinatubo.

Figure 14.14 (bottom) compares the trends of mean zonal temperature in simulations and observations, with the 2 years following El Chichón and Pinatubo excluded. There is cooling in all latitudes in low- and high-top ensemble means and also in observations. There is no significant difference between the degree of agreement with observations of the low- and high-top simulations, and the level of

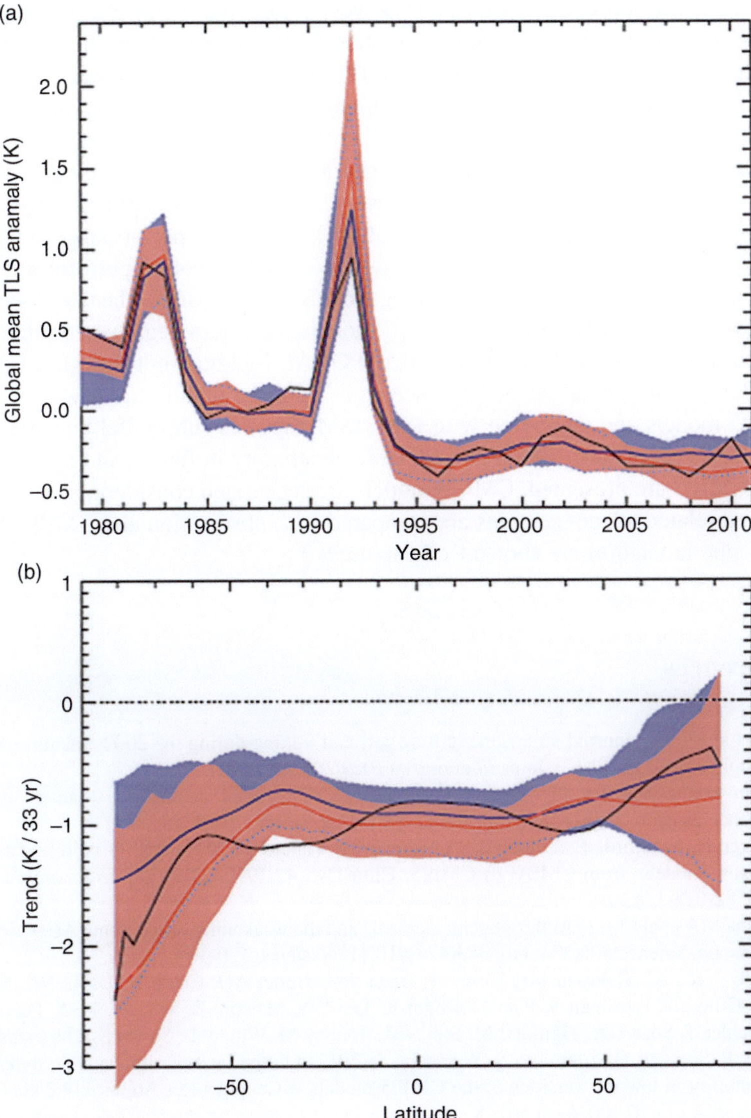

Fig. 14.14 Comparison of observed and simulated temperature in lower stratosphere. (**a**) The average for the low-top models (blue), the average of the high top (red), global annual mean lower stratospheric temperature anomalies from the 1979–2011 mean, based on the dataset from remote sensing solutions (black). Pink and blue shaded bands mark the ~5–95% ensemble range. Global means are calculated over the latitude belt 82.5°N–82.5°S. (**b**) Corresponding zonal mean temperature trends (°C) over 1979–2011. Linear trends were fitted by least squares, and volcano years were excluded (1992, 1993, 1991, 1982 and 1983) from the analysis (Charlton-Perez et al. 2013)

agreement with observations is good. However, observed trends during the period 1979–2005 indicate little latitudinal variation (Gillett et al. 2011). This could be in part due to anomalous hot Antarctic in 2002 (Allen et al. 2003; Varotsos 2002). The trend calculated over the period 1979–2011 suggests enhanced cooling in Antarctic, which is consistent with the simulations. Overall, the CMIP5 models considered can simulate changes in the lower stratospheric observed temperature rather well.

In this chapter, we discussed the following. Model outputs in global temperature are presented and compared with observation. For ISM, the performance of CMIP5 models is studied. Regarding climatology, some CMIP5 models perform well. But some features of ISM vary widely among models and discussed. The teleconnection between ISM and ENSO is studied. All models indicate a negative correlation in historical as well as RCP scenario. Model FGOALS-g2 does not show any trend in Niño3.4 for the historical or RCP scenario. Global rainfall was analysed using CMIP5 models and compared with ISM. ISM suggests substantial uncertainty in future. The ENSO also shows considerable uncertainty in future, and some results from CMIP3 are presented. CMIP5 models are segregated considering low or high top, and related teleconnections are compared with observation and AMIP5. Some stratospheric features are shown well captured.

References

Allen et al (2003) Unusual stratospheric transport and mixing during the 2002 Antarctic winter. Geophys Res Lett 30(12). https://doi.org/10.1029/2003GL017117

Baldwin MP, Dunkerton TJ (2001) Stratospheric harbingers of anomalous weather regimes. Science 294(5542):581–584. https://doi.org/10.1126/science.1063315

Bellenger H, Guilyardi E, Leloup J, Lengaigne M, Vialard J (2014) ENSO representation in climate models: from CMIP3 to CMIP5. Clim Dyn 42:1999–2018. https://doi.org/10.1007/s00382-013-1783-z

Bollasina MA et al (2011) Anthropogenic aerosols and the weakening of the South Asian summer monsoon. Science 502:334. https://doi.org/10.1126/science.1204994

Charlton-Perez AJ, Baldwin MP, Birner T, Black RX, Butler AH, Calvo N, Davis NA, Gerber EP, Gillett N, Hardman S, Kim J, Krüger K, Lee Y-Y, Manzini E, McDaniel BA, Polvani L, Reichler T, Shaw TA, Sigmond M, Son S-W, Toohey M, Wilcox L, Yoden S, Christiansen B, Lott F, Shindell D, Yukimoto S, Watanabe S (2013) On the lack of stratospheric dynamical variability in low-top versions of the CMIP5 models. J Geophys Res Atmos 118:2494–2505. https://doi.org/10.1002/jgrd.50125

Collins M et al (2010) The impact of global warming on the tropical Pacific Ocean and El Niño. Nat Geo Sci:391. https://doi.org/10.1038/NGEO868

Gillett NP et al (2011) Attribution of observed changes in stratospheric ozone and temperature. Atmos Chem Phys 11:599–609

Hurrell JW, Hack JJ, Shea D, Caron JM, Rosinski J (2008) A New Sea surface temperature and sea ice boundary dataset for the community atmosphere model. J Clim 21:5145–5153. https://doi.org/10.1175/2008JCLI2292.1

Hurwitz MM, Calvo N, Garfinkel CI, Butler AH, Ineson S, Cagnazzo C, Manzini E, Peña-Ortiz C (2014) Extra-tropical atmospheric response to ENSO in the CMIP5 models. Clim Dyn 43:3367–3376. https://doi.org/10.1007/s00382-014-2110-z

Jourdain NC, Sen Gupta A, Taschetto AS, Ummenhofer CC, Moise AF, Ashok K (2013) The Indo-Australian monsoon and its relationship to ENSO and IOD in reanalysis data and the CMIP3/CMIP5 simulations. Clim Dyn 41:3073–3102. https://doi.org/10.1007/s00382-013-1676-1

Maity R, Kumar DN (2006) Bayesian dynamic modelling for monthly Indian summer monsoon rainfall using El Niño–Southern Oscillation (ENSO) and Equatorial Indian Ocean Oscillation (EQUINOO). J Geophys Res 111:D07104

Roy I (2017) Indian summer monsoon and El Niño southern oscillation in CMIP5 models: a few areas of agreement and disagreement. Atmos 8:154

Roy I, Tedeschi RG (2016) Influence of ENSO on regional ISM precipitation – local atmospheric Influences or remote influence from Pacific Atmosphere 7:25 https://doi.org/10. 3390/atmos7020025

Roy I, Tedeschi RG, Collins M (2017) ENSO Teleconnections to the Indian summer monsoon in observations and models. Int J Climatol. https://doi.org/10.1002/joc.4811

Taylor KE, Stouffer RJ, Meehl GA (2012) An overview of CMIP5 and the experiment design. Bull Amer Meteor Soc 93:485–498. https://doi.org/10.1175/BAMS-D-11-00094.1

Turner AG, Inness PM, Slingo JM (2005) The role of the basic state in the ENSO-monsoon relationship and implications for predictability. Q J R Meteorol Soc 131(607):781–804

Varotsos (2002) The southern hemisphere ozone hole split in 2002, environtal. Sci Pollut Res 9(6):375–376

Chapter 15
Green House Gas Warming

Abstract This chapter focused on greenhouse effect and covered various relevant areas. Starting from laws of radiation, it included discussion on solar radiation vs. terrestrial radiation, atmospheric windows for terrestrial radiation, water vapour and carbon dioxide (CO_2) as greenhouse gases, radiative forcing and global energy balance, etc.

Keywords Green House Gas · CO_2 · Water vapour · Wien's Displacement Law · Global energy balance · Radiative forcing

To discuss greenhouse gas warming, let us start with famous laws of radiation.

15.1 Laws of Radiation

Among laws of radiation, Wien's displacement law plays a major role in describing greenhouse gas warming. It states maximum energy output in a wavelength (λmax) is inversely proportional to temperature T ($^\circ$K) of the body and can be written as

$$\lambda\text{max} \times T = 2880, \ \lambda\text{max is in micron}$$

If $T = 288$ K and λmax $= 10$ micron, it is true for the Earth.
And if $T = 6000$ K and λmax $= 2880/6000 = 0.48$ micron, it applies to the Sun.

$$1\,\text{micron} = 10^{-6}\,\text{m}$$

15.2 Solar Radiation vs. Terrestrial Radiation

From Wien's displacement law, it is seen the Sun emits radiation with the peak intensity near 0.48 micron, and the human eye is sensitive to this radiation. Violet with wavelength 0.4 micron and red 0.7 micron are two ends of the range. Hence

© Springer International Publishing AG, part of Springer Nature 2018
I. Roy, *Climate Variability and Sunspot Activity*, Springer Atmospheric
Sciences, https://doi.org/10.1007/978-3-319-77107-6_15

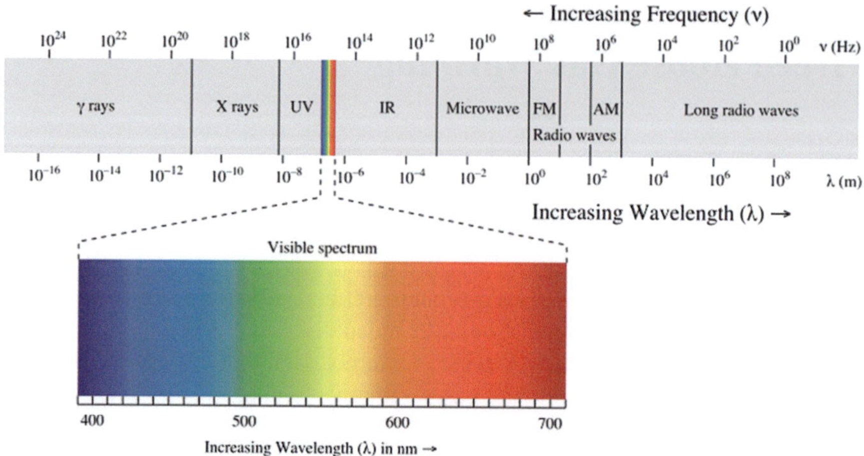

Fig. 15.1 The wavelength of different radiation from decreasing to increasing order with respective frequencies. Visible range is shown in an enlarged schematic at the bottom. (Source: http://en.wikipedia.org/wiki/File:EM_spectrum.svg, uploaded on 30/12/17)

electromagnetic waves within this range are called visible, and ~50% of the solar radiation lies within this range. A major portion of the solar radiation among the rest are ultraviolet (UV) (0.01–0.4 micron) and infrared (IR) (0.7–100 micron). Radiation shorter than UV and larger than IR is also present, but that amount is insignificant. The wavelength of various radiations with respective frequencies is shown in Fig. 15.1.

Radiation emitted by the Earth is known as terrestrial radiation. From Wien's displacement law, the Earth emits radiation with the peak intensity near 10 microns. As this is much longer than that of the Sun, it is known as the longwave in contrast to the shortwave radiation of the Sun.

15.3   Radiation Transmitted by the Atmosphere and Atmospheric Windows

Figure 15.2 top panel shows radiation transmitted by the Sun (top left) and the Earth (top right). Various components those mainly absorb solar and terrestrial radiation are shown in the bottom panel. The percentage of total absorption and scattering by different components are presented in the middle panel.

15.4   Absorption: Water Vapour and CO_2

Water vapour and CO_2 strongly absorb infrared radiation, and lower atmosphere is mainly heated this way. Both water vapour and CO_2 are transparent to infrared radiation of bandwidth 7–12 micron (shown in Fig. 15.3) which includes peak intensity

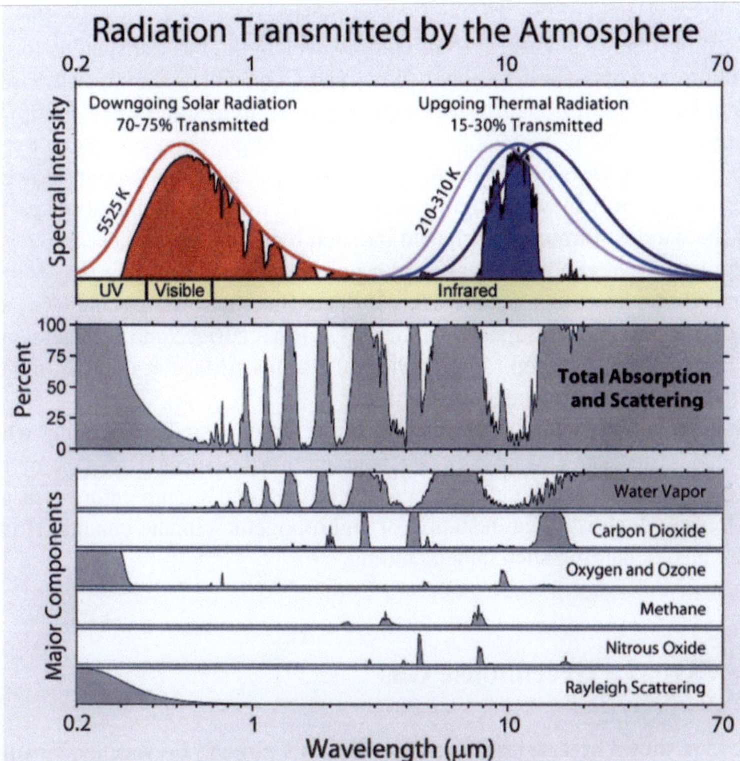

Fig. 15.2 Radiation transmitted by the sun (red) and earth (blue) [top]. Various components those mainly absorb solar and terrestrial radiation are shown [bottom] with respective wavelength ranges. Total percentage of absorption and scattering in different wavelengths are in the middle panel. (Source: https://commons.wikimedia.org/wiki/File:Atmospheric_Transmission.png. CC BY-SA 3.0, image loaded on 30/12/2017)

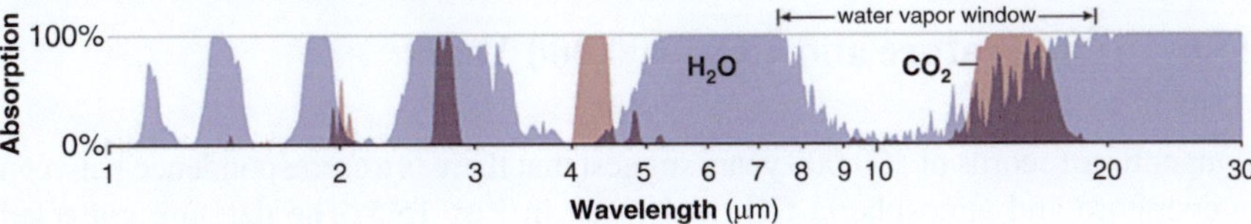

Fig. 15.3 Absorption of infrared radiation in different wavelength range by CO$_2$ (violet) and H$_2$O (blue). (Source: earthobservatory.nasa.gov/Features/EnergyBalance/page7.php. uploaded on 30/12/2017, credit NASA's Earth Observatory)

of terrestrial radiation (also shown in the bottom panel of Fig. 15.2). This bandwidth is called atmospheric window or infrared window, and the Earth loses heat through this window and get cooled.

There are other atmospheric windows such as Radio-window of wavelength 10 mm to ~80 m in radio wave range. TV transmitters, radio and radars are designed to operate in this region.

The bottom panel of Fig. 15.2 suggests atmospheric gases have a specific pattern of absorbing energy – absorbing in some wavelengths but transparent to others. Absorption patterns of water vapour (blue) and CO_2 (violet) are superimposed and shown in Fig. 15.3. It is clearly seen that they overlap in some wavelength ranges. As is seen there, that water vapour absorption is in larger range compared to CO_2 in IR band. Hence CO_2 is not as strong greenhouse gas as water vapour. However, it absorbs more energy in wavelength bands of 12–15 μm. It partially closes particular part of the window through which heat radiated by surface would escape.

Definition of greenhouse effect: Major portion of solar radiation passes through the atmospheric gases and heats Earth's surface. But water vapour and CO_2, absorb infrared (IR) part of the radiation emitted by Earth's surface and reradiates it back to the Earth. It is the reason rapid cooling of Earth's surface at night is prevented. This process is known as greenhouse effect.

Due to increased industrialisation, CO_2 of the atmosphere is increasing, which in turn impedes the escape of IR radiation from the Earth's surface to space by greenhouse effect. Scientists, therefore, apprehend that Earth's temperature will gradually increase which is a manifestation of anthropogenic climate change. However, there are also debates in such understanding.

15.5 CO_2 as a Greenhouse Gas

Figure 15.4 shows greenhouse forcing on Earth's climate by various constituents (top). As is seen, water vapour is the major greenhouse gas which constitutes nearly half of the overall forcing. The bottom panel of Fig. 15.4 shows major greenhouse gases from people's activity, where CO_2 dominates. In this category, CO_2 is followed by methane.

15.6   Temperature and CO_2: 400,000 Years

Longer time records of 400,000 years suggest that there is a correspondence between temperature and atmospheric CO_2 as is seen in Fig. 15.5. The data are collected from Antarctic ice-core records.

15.7   Earth's Temperature Change in the Last 2000 Years

Figure 15.6 shows reconstructions of surface temperature changes from various research groups. The range of uncertainty increases backwards in time. As is seen in Fig. 15.6, there is reasonable agreement among various reconstructions and instrumental record since ~1850.

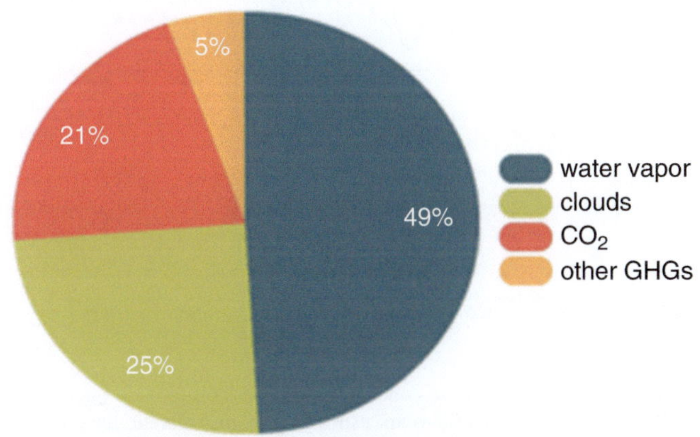

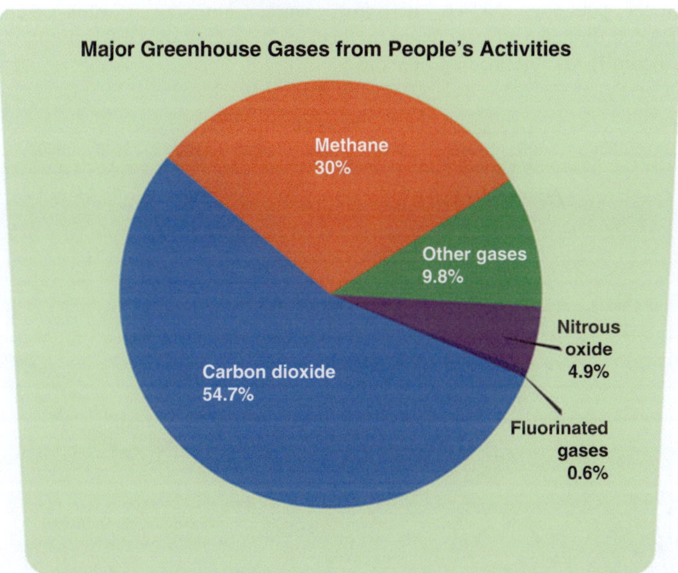

Fig. 15.4 The top panel shows greenhouse forcing on Earth's climate by various constituents; bottom panel shows major greenhouse gases from people's activity. (Source: Carbon Connections. http://carbonconnections.bscs.org. Copyright © BSCS. All rights reserved. Used with permission)

15.8   Radiative Forcing

Radiative forcing (RF) is defined as the change in net radiative flux measured at the tropopause level. It is done after allowing for stratospheric temperatures to readjust to radiative equilibrium but with tropospheric and surface temperatures held constant.

A fundamental concept in the area of climate change is that of 'radiative forcing' which is loosely defined as the instantaneous net radiative imbalance that takes

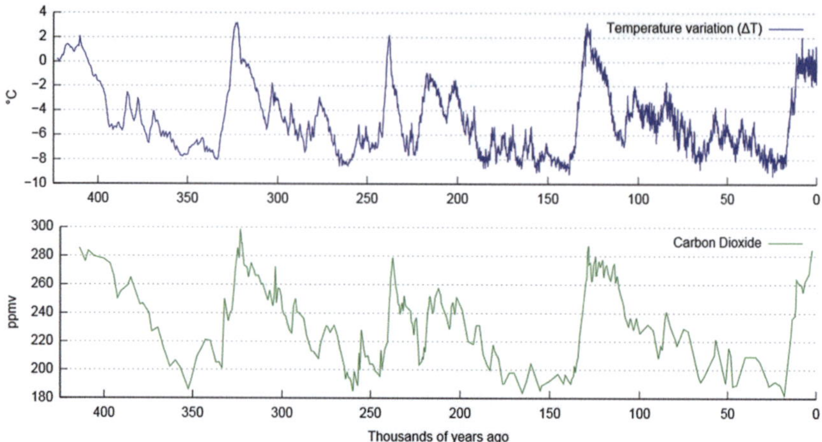

Fig. 15.5 Fluctuations in atmospheric concentration of carbon dioxide (green) and temperature (shown in blue) over the last 400,000 years using data from Antarctic ice-core records. Source: https://commons.wikimedia.org/w/index.php?curid=10684392, loaded on 19/12/2017. (By Vostok-ice-core-petit.png: NOAA derivative work: Autopilot (Vostok-ice-core-petit.png) [CC-BY-SA-3.0 (http://creativecommons.org/licenses/by-sa/3.0/) or GFDL (http://www.gnu.org/copyleft/fdl.html)], via Wikimedia Commons.)

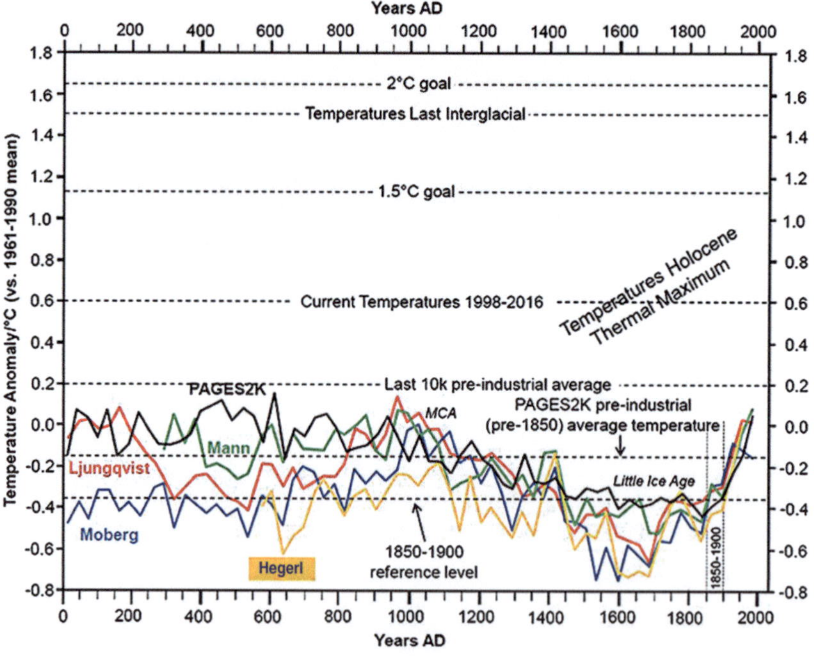

Fig. 15.6 Temperature reconstructions for the past 2000 years by various research teams (Moberg, Hegerl, Mann, Ljungqvist, and PAGES 2 k Consortium, shown in different colour shades) compared to important reference levels. Average pre-industrial temperatures for the last 10,000 years are shown by 'last 10 k', whereas 'MCA' denotes Medieval Climate Anomaly. The range of uncertainties increases backwards. (Source: https://www.frontiersin.org/files/Articles/317793/feart-05-00104-HTML/image_m/feart-05-00104-g002.jpg, uploaded on 19/12/2017, Copyright © 2017 Lüning and Vahrenholt)

place at the top level of the atmosphere introduced by some perturbing factor. The latter could be a change in atmospheric composition or incoming solar irradiance or of the Earth's albedo. Radiative forcing is a useful concept as it gives a top level indication of the potential impact of the perturbing factor on the equilibrated global mean surface temperature of the Earth. Because of this, it is used extensively to indicate the relative effects of various anthropogenic and natural factors on climate.

Figure 15.7 shows radiative forcing since 1750 as set out in the Fourth Assessment Report of the Intergovernmental Panel on Climate Change (IPCC 2007), while Fig. 15.8 is as defined in the Fifth IPCC Report (IPCC 2013). The solar contribution in both the figures is perceived to be very small, but the ascribed 'Level of Scientific Understanding' is low. In this context, it should be noted this value takes no account of either direct or indirect feedback factors on solar radiative forcing (such as due to changes in ozone or to cloud cover) nor can it be used to interpret geographical distributions of any solar impact.

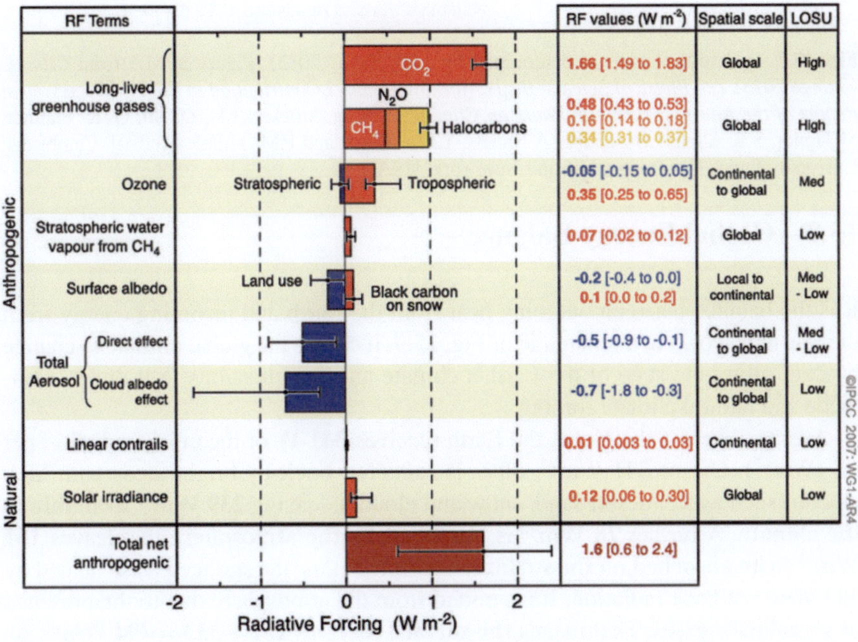

Fig. 15.7 Components of radiative forcing 1750–2005 as defined by IPCC (2007) (Figure SPM.2 from *Climate Change 2007: The Physical Science Basis. Working Group I Contribution to the Fourth Assessment Report of the Intergovernmental Panel on Climate Change* [Solomon, S., D. Qin, M. Manning, Z. Chen, M. Marquis, K.B. Averyt, M. Tignor and H.L. Miller (eds.)]. Cambridge University Press, Cambridge, United Kingdom and New York, NY, USA.)

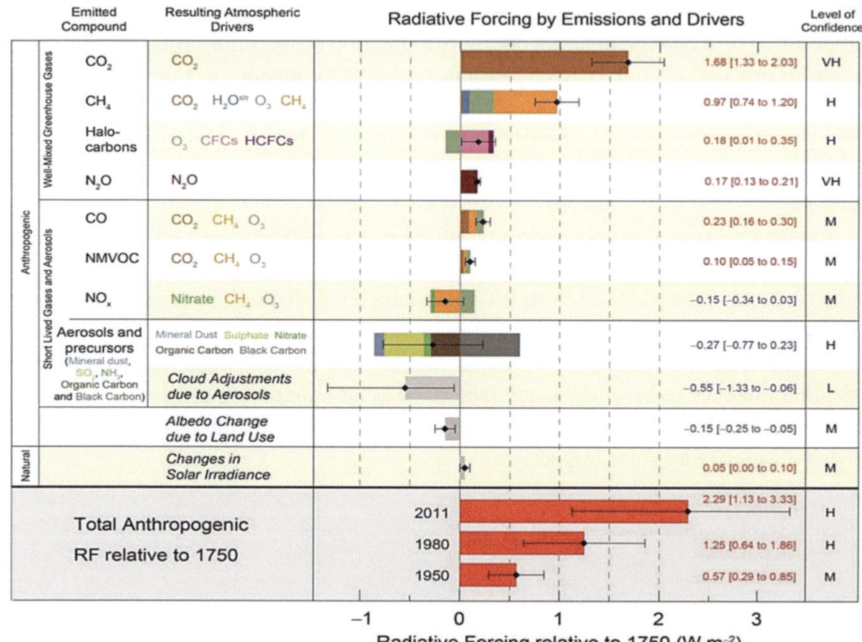

Fig. 15.8 Radiative forcing as determined by AR5 (IPCC 2013) (Figure SPM.5 from *Climate Change 2013: The Physical Science Basis. Working Group I Contribution to the Fifth Assessment Report of the Intergovernmental Panel on Climate Change* [Stocker, T.F., D.Qin, G.-K. Plattner, M.Tignor, S.K.Allen, J.Boschung, A.Nauels, Y.Xia, V.Bex and P.M. Midgley (eds.)]. Cambridge University Press, Cambridge, UK and New York, USA.)

15.9 Global Energy Balance

It is the balance between outgoing heat from the Earth and incoming energy from the Sun and shown in a schematic of Fig. 15.9. It causes the global climate to change by controlling the state of the Earth's climate and modifications to it due to man-made and natural climate forcing.

Averaged across the globe, the Earth receives **341 W** of incoming radiation per m² (Wm⁻²). Out of 341 Wm⁻², **30% is reflected back** by bright areas with high albedo (such as ice, desert sand, snow and clouds), leaving **239 Wm⁻² available to the climate**. A further **78 Wm⁻² is absorbed by the atmosphere** that leaves **161 Wm⁻² to be absorbed on the surface.** Apart from this, the surface is also heated by 333 Wm⁻² of back radiation. It is emitted from the atmosphere due to the presence of greenhouse gases. That means **the surface receives 161 + 333 = 494 Wm⁻².** As energy cannot be destroyed or created, the Earth needs to emit 494 Wm⁻² back from the surface to keep the energy balance in equilibrium. A very small amount is emitted through evapotranspiration and thermal transfer (80 and 17 Wm⁻², respectively), and the rest 396 Wm⁻² is emitted as heat radiation by the surface. However, **only**

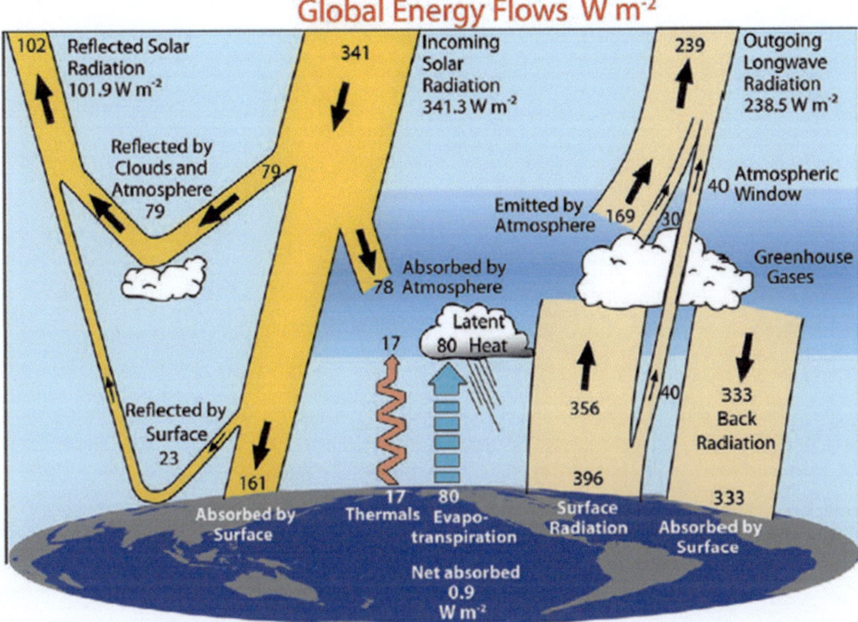

Fig. 15.9 The annual global mean Earth's budget of energy in W m^{-2} [March 2000 to May 2004]. The thick arrows suggest the schematic energy flow in proportion to their importance (Trenberth et al. 2009)

239 Wm^{-2} leaves the top of the atmosphere. It is because of trapping by greenhouse gases and equals to the initial amount of solar radiation absorbed.

References

IPCC (2007) Fourth assessment report of the intergovernmental panel on climate change. Cambridge University Press, Cambridge/New York

IPCC, Climate Change (2013) The physical science basis. Contribution of working group I to the fifth assessment report of the intergovernmental panel on climate change. Cambridge University Press, Cambridge/New York. http://www.climatechange2013.org/images/report/WG1AR5_ALL_FINAL.pdf

Trenberth et al (2009) Earth's global energy budget. Bull Amer Meteorol Soc 90:311–324. https://doi.org/10.1175/2008BAMS2634, pp311

Chapter 16
Volcanic Influences

Abstract Volcano causes major changes in the Earth's atmosphere, and hence this chapter covers various influences it inflicts to the climate of Earth that involve both the stratosphere and troposphere. The different time scales of influence are also covered. Names of major volcanos in the last 250 years are mentioned. Few other areas also addressed are polar warming associated with large eruption and the combined influence of the Sun, volcano and ENSO.

Keywords Volcano · El Chichon · Pinatubo

Volcanic eruptions play major influences in the weather and climate of the earth. Table 16.1 shows the years of major volcanos of the last 250 years with the country of origin. The last two major volcanos El Chichon, in 1982, and Mount Pinatubo, in 1991 are most powerful eruptions in the last century, and they were associated with various effects on the climate of the earth in different time scales.

16.1 Volcano Cooling Effect

Volcanos are found to produce cooling effect throughout the years lasting for the next 2–3 years time. Figure 16.1 shows average temperature anomaly for five volcanic eruptions from the last century. Cooling is seen even after the fourth year, and the maximum cooling is noticed during the first and second year that even reaches the magnitude of −0.2 °C.

Mount Pinatubo in 1991 was one of the century's most powerful eruptions. Its massive dust and aerosol cloud cooled parts of the world by up to 0.4 °C.

© Springer International Publishing AG, part of Springer Nature 2018
I. Roy, *Climate Variability and Sunspot Activity*, Springer Atmospheric
Sciences, https://doi.org/10.1007/978-3-319-77107-6_16

Table 16.1 Major volcanos of the last 250 years

Volcano	Eruption year
Grimsvotn (Lakagigar) in Iceland	1783
Tambora in Sumbawa, Indonesia	1815
Cosiguina in Nicaragua	1835
Askja in Iceland	1875
Krakatau in Indonesia	1883
Okataina (Tarawera) in North Island, New Zealand	1886
Santa Maria in Guatemala	1902
Ksudach in Kamchatka, Russia	1907
Novarupta (Katmai) in Alaska, United States	1912
Agung in Bali, Indonesia	1963
Mount St. Helens in Washington, United States	1980
El Chichon in Chiapas, Mexico	1982
Mount Pinatubo in Luzon, Philippines	1991

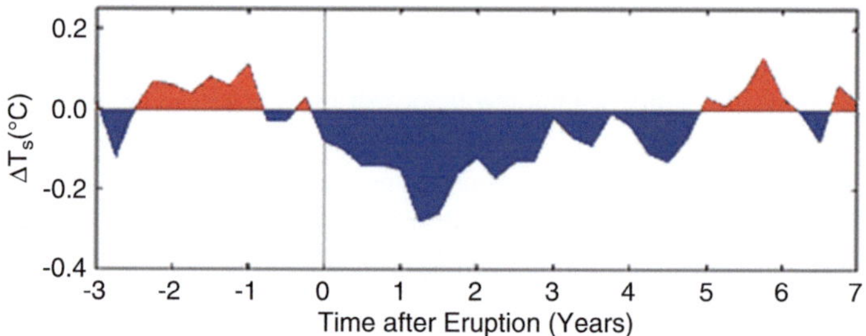

Fig. 16.1 The average effect of the five strongest volcanic eruptions this century is a cooling impact up to 0.2 degrees C, lasting for about 2 years. (Source: NASA-GISS, http://www.carbonbrief.org/in-brief-how-much-do-volcanoes-influence-the-climate uploaded on 30/12/17, used with permission)

16.2   Influences of Volcanic Eruption

Volcanic eruptions can inject enormous amounts of solid particles (ash) and gases into the upper atmosphere. Due to its size and mass, volcanic ash rapidly falls out of atmosphere again. Therefore, volcanos' climate response mostly results from emission of sulphurous gases and resulting aerosol particles. Their concentration can exceed background stratospheric aerosol by several magnitudes. Stratospheric aerosols influence global climate system in many ways. By scattering incoming solar radiation and absorbing Earth's thermal radiation, they directly affect radiation (shown in Fig. 16.2).

As a consequence, aerosol-containing layers in stratosphere warm up, whereas near-ground air layers and ocean cool down. Moreover, heterogenic chemical reactions occur on the surface of volcanic aerosol particles leading to activation of chlorine and thus reduction of ozone.

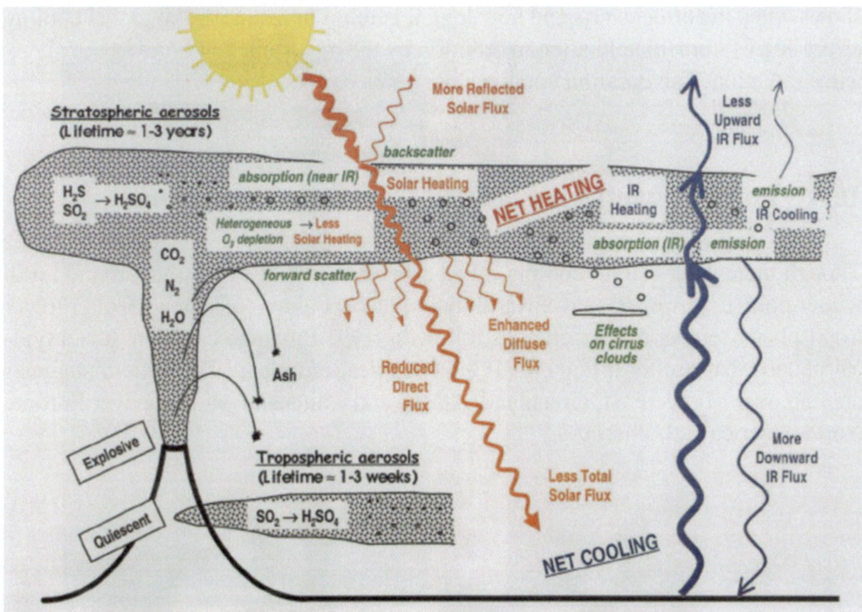

Fig. 16.2 Volcanic influences in the troposphere and stratosphere (Robock 2000)

Table 16.2 Effect of large volcanic eruptions on climate and weather (Robock 2000)

Effect	Mechanism	Begins	Duration
Reduced diurnal cycle	Emission of longwave and blockage of shortwave radiation	Immediately	1–4 days
Reduced tropical precipitation	Emission of longwave and blockage of shortwave radiation	1–3 months	3–6 months
Summer cooling of NH tropics and subtropics	Blockage of shortwave radiation	1–3 months	1–2 years
Warming in stratosphere	Absorption of shortwave and longwave radiation in stratosphere	1 to 3 months	1–2 years
Winter warming of NH continents	Absorption of shortwave and longwave radiation in stratosphere, dynamics	Half year	1–2 winters
Cooling in globe	Shortwave radiation blocking	Immediately	1–3 years
Cooling in globe from multiple eruptions	Shortwave radiation blocking	Immediately	10–100 years
Enhanced UV and ozone depletion	Heterogeneous chemistry on aerosols and dilution	1 day	1–2 Years

16.3 Effect of Large Eruptions on Weather and Climate

Robock (2000) discussed the time scale of the consequences of the massive volcanic eruptions on various weather and climate patterns and are presented here in Table 16.2. Different effects are mentioned with the mechanism involved. It also shows when the effect starts and how long it is felt. For example, the global cooling effect begins immediately after an eruption by the mechanism of blockage of short-wave radiation. The duration could be 1–3 years period.

16.4 Polar Warming Associated with Large Eruptions

Though there is an overall cooling effect after the massive explosion, around north winter pole, it, however, shows warming in places. Figure 16.3 shows winter lower tropospheric temperature anomalies following 1991 Pinatubo eruption. It is a typical pattern that occurs following all large tropical eruptions. The pattern suggests cooling over Middle East, Greenland, Alaska and China and warming over Europe, North America and Siberia.

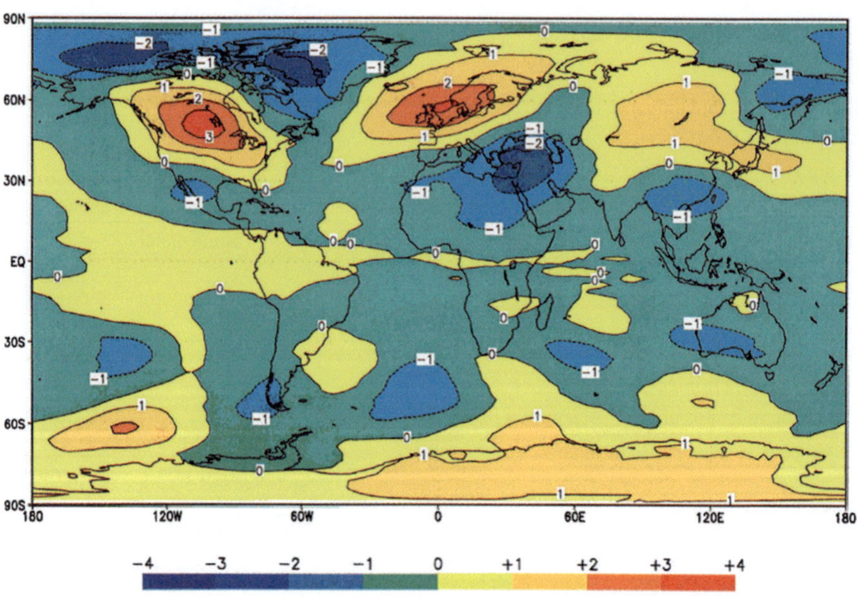

Fig. 16.3 Winter lower tropospheric temperature (DJF, 1991–1992) anomalies in °C using data from microwave sounding unit channel 2R. To calculate the mean for 1991–1992 NH winter (DJF) following 1991 Pinatubo eruption, a non-volcanic period of 1984–1990 is considered (Robock 2000)

16.5   Sun, Volcano and ENSO

Figure 16.4 is a time series plot of various independent factors that are likely to influence the climate of the Earth. Here linear trend is not shown. The first plot on top is for volcanic eruptions, which is represented by stratospheric aerosol optical depth (OD). The middle plot is the time series of solar 11-year cycle variability (here SSN), whereas the bottom one is to that for ENSO.

Not only are the strength of eruption and the power of solar cycle necessary but also their combined behaviour, which includes the timing of eruption, regarding the phase of the solar cycle, are important in controlling the climate of the earth. The period during 1917 –1944 was a period of very little volcanic activity that coincided with an increase in solar irradiance (Fig. 16.4, marked by II). The late 1800s to the early 1900s was again a very active period of volcanic eruptions, though the Sun

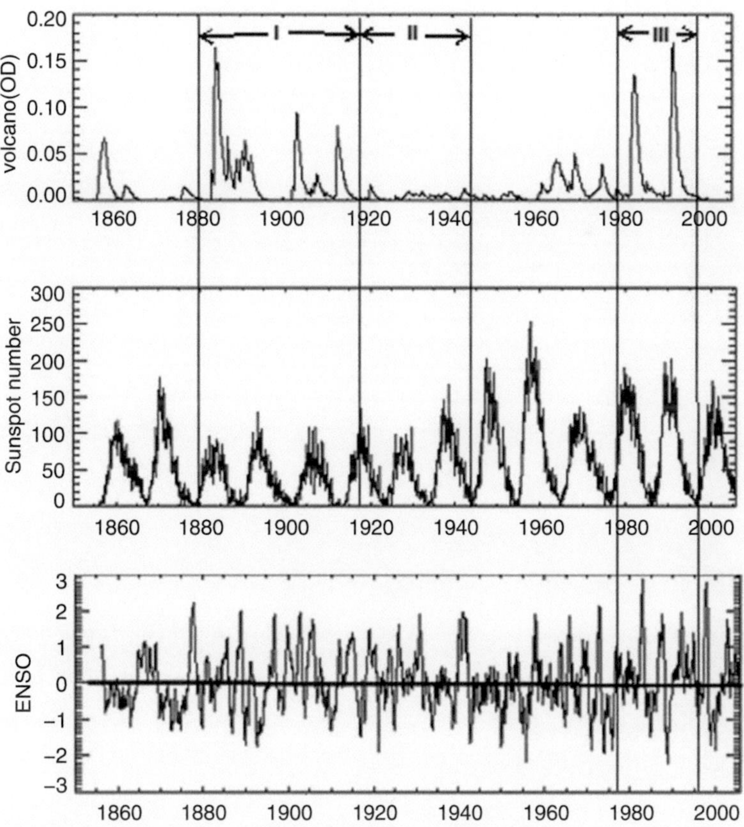

Fig. 16.4 Time series of the volcano, SSN and ENSO (Roy 2016)

was quiet (marked by I, Fig. 16.4). During the latter half of the last century, the period dominated by climate change, the solar cycles are also seen to be stronger, as noticed in Fig. 16.4. Two very active volcanos erupted during near peak of active solar cycles (Fig. 16.4, period III). During period III, ENSO is also seen to be stronger and more variable.

References

Robock (2000) Volcanic Eruptions and Climate. Rev Geophys 38(2):191–219
Roy I (2016) The role of natural factors on major climate variability in Northern Winter. Preprints 2016, 2016080025 https://doi.org/10.20944/preprints201608.0025.v2

Chapter 17
Ozone Depletion in the Stratosphere

Abstract This chapter discusses Antarctic ozone hole and its relevance to banning of CFCs through Montreal Protocol. It describes that ozone in the stratosphere and greenhouse gas in the troposphere both have a comparable effect on surface climate, though ozone is more important in higher latitudes and altitudes.

Keywords Ozone hole · Montreal Protocol · UV radiation · Anthropogenic influence

Climate change during recent decades is mainly attributed to the collective anthropogenic influences of decreasing stratospheric ozone and increasing greenhouse gases. Reduction of stratospheric ozone and formation of the ozone hole are playing an important part and hence discussed here.

17.1   Ozone Hole and Montreal Protocol

Ozone by absorbing UV radiation from the Sun protects living organisms underneath from its harmful radiation. The depletion of this protective ozone layer over Earth's polar region is referred as 'ozone hole'. In the 1980s, scientists discovered ozone layer was thinning in the lower stratosphere, with particularly dramatic ozone loss causing 'ozone hole' in Antarctic springtime (September and October). It is due to increase concentrations of ozone-depleting chemicals in the stratosphere that come from refrigerants and spray cans. These long-lived CFCs (chemical compounds with chlorine and fluorine attached to carbon) or chlorofluorocarbons remain in the atmosphere for decades. At poles, CFCs attach to ice particles. When the sun comes out in polar spring, ice particles melt, releasing ozone-depleting molecules. They break molecular bonds in UV radiation-absorbing ozone.

An international treaty is signed on 1987 named as **Montreal Protocol on Substances that Deplete the Ozone Layer.** The purpose is to phase out production of substances those are responsible for ozone depletion and thus to protect the ozone layer.

© Springer International Publishing AG, part of Springer Nature 2018

I. Roy, *Climate Variability and Sunspot Activity*, Springer Atmospheric Sciences, https://doi.org/10.1007/978-3-319-77107-6_17

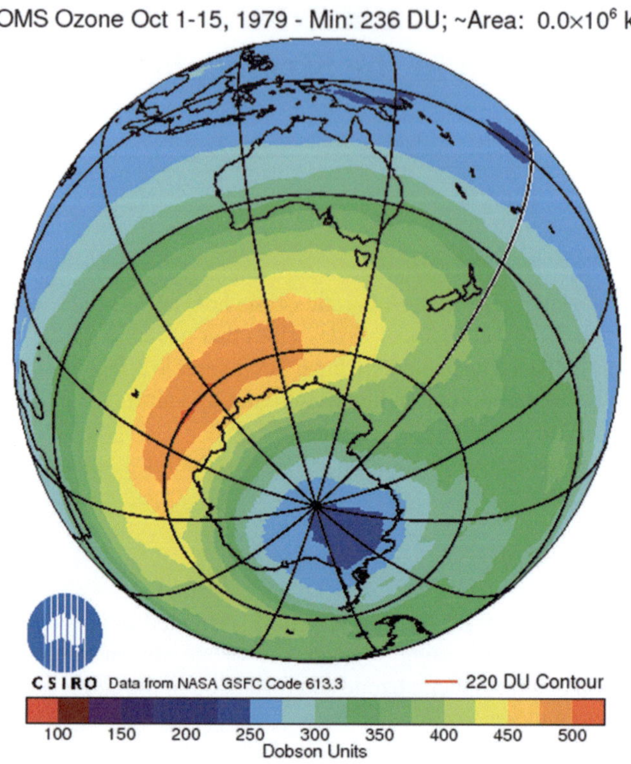

Fig. 17.1 (Animation): Amount of ozone around mid-southern latitude to the pole in Dobson Unit (DU) is shown by colours. (Source: http://www.environment.gov.au/protection/ozone/ozone-science/ozone-layer/antarctic-ozone-hole, Animation link (dt 30/12/17))

17.2   Ozone Hole Animation

The amount of ozone around mid-southern latitude to the pole during October (1–15) is shown in different years from 1979 to 2010 in Fig. 17.1 (animation). The ozone hole is formed each year in the southern hemisphere spring when there is a sharp decline (currently up to 60%) in the total ozone over most of the Antarctica. The hole marked by red around Antarctica as seen in Fig. 17.1 (animation of later years) is often termed as ozone hole, and it varies in size and shape year to year.

17.3   Greenhouse Gases and Ozone in Model

Effect of greenhouse gases and ozone loss is studied by Shindell et al. (2004) in models and presented in Fig. 17.2. Temperature is shown by colours and zonal wind by solid contours. Their model precisely captures greenhouse warming in the

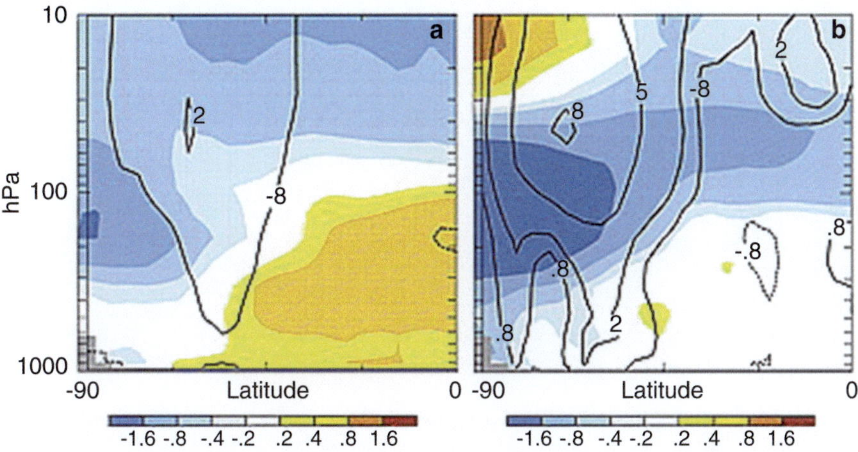

Fig. 17.2 Zonal mean Dec–May temperature (in °C, shown with colours) and zonal wind (m/s, shown with contours) linear trends in the simulations from 1970 to 1999 forced by rising greenhouse gases (**a**) and by ozone depletion (**b**) (Shindell et al. 2004)

troposphere. However, they showed, an influence of ozone loss is clearly paramount in the lower stratosphere, leading, e.g. to a zonal wind anomaly nearly three times greater than that caused by greenhouse gases during the last three decades of the twentieth century. Overall, the effect of ozone becomes more necessary for higher latitudes and altitudes, though extratropic wide SLP signals of ozone and greenhouse gases are not statistically different. According to Shindell et al. (2004), two factors had comparable impacts on surface climate during that period, though above the middle troposphere it is dominated by ozone.

Reference

Shindell D et al (2004) Southern Hemisphere climate response to ozone changes and greenhouse gas increases. Geophys Res Lett 31:L18209. https://doi.org/10.1029/2004GL020724

Chapter 18
Influence of Various Other Solar Outputs

Abstract Apart from solar 11-year cyclic variability, other solar outputs are also important in modulating the climate of the Earth and discussed briefly in this chapter. It discussed various routes through which solar variability may influence the climate of the lower atmosphere. At the end, there are also brief discussions on galactic cosmic ray.

Keywords TSI · UV flux · Galactic Cosmic ray · Energetic electron precipitation (EEP)

Though regarding solar output, we mainly discussed solar output relating to UV variability, but other solar outputs like cosmic ray, energetic particle, etc. are also necessary, and their region of influence is shown in Fig. 18.1.

Because of close relationship between SSN and various solar radiation-related drivers (TSI and UV flux), SSN may be served as a proxy for analysing sun-climate connection (Roy 2014, 2018; Roy and Collins 2015). Whereas, the solar cycle evolution of geomagnetic activity, solar wind speed and energetic electron precipitation (EEP) has been different from SSN at least during the space age. Those peaked in the declining phase of SSN cycles.

Several studies found significant association between surface winter climate and geomagnetic activity (Maliniemi et al. 2014; Bochnıcek et al. 2012; Roy et al. 2016). Geomagnetic activity is in fact solar wind-related driver and shown to impact surface climate (Bochnıcek and Hejda 2005; Bochnıcek et al. 2012)). It has two separate components, the first one due to coronal mass ejections (CME), while the second arises from high-speed solar wind streams (HSS). CME roughly shows an in-phase variation with the SSN, whereas the HSS have a variable lag with SSN cycle, with its peak in the declining phase of the SSN cycle.

Energetic particles are responsible for ionisation of the upper/middle atmosphere and thus have an influence on the atmospheric temperature and composition. Those particles are important for the coupling in magnetosphere, ionosphere and thermosphere. Through NOx production those impact the lower polar stratosphere and act as a potential route to modulate NAO (Maliniemi et al. (2013)). A possible mechanism was proposed that linked winter circulation variability at polar regions

© Springer International Publishing AG, part of Springer Nature 2018

I. Roy, *Climate Variability and Sunspot Activity*, Springer Atmospheric
Sciences, https://doi.org/10.1007/978-3-319-77107-6_18

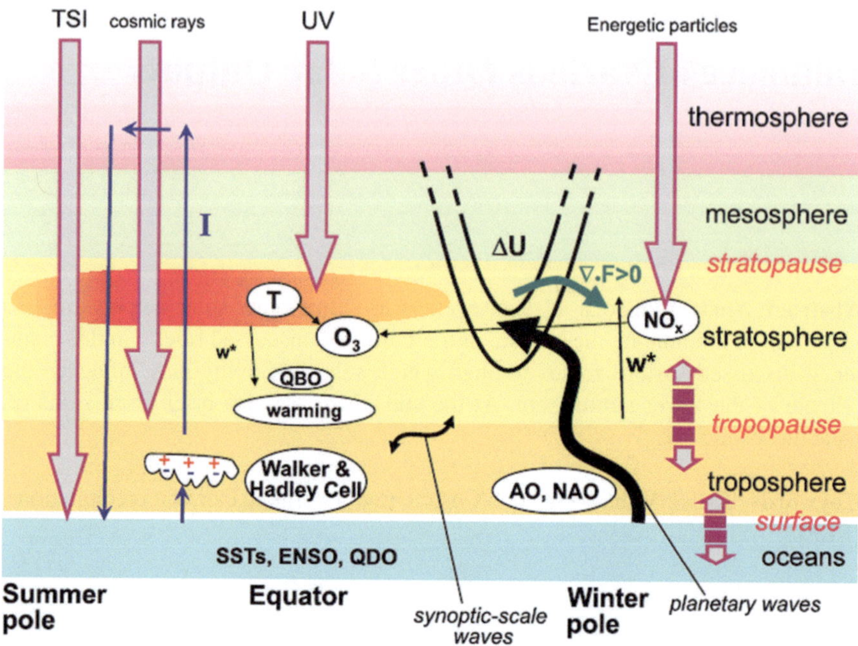

Fig. 18.1 Various solar outputs, e.g. TSI, cosmic rays, UV and energetic particles, are shown with the region of their influences (Gray et al. (2010))

involving energetic particle precipitation. It is initiated via changes in atmospheric chemistry, e.g. by producing nitrogen oxides (NOx) in the upper atmosphere (Seppälä et al. 2007; Sinnhuber et al. 2011). NOx can descend down (Funke et al. 2005) and affect ozone balance in the stratosphere during polar winter. In winter, NOx have long lifetimes because of the absence of sunlight and a large-scale downward propagation prevails at poles. Such mechanism is also noticed in the chemistry-climate models (Rozanov et al. 2005; Baumgaertner et al. 2011).

18.1   Mechanisms

Though a solar signal is evident in many climate parameters, and it is obvious that the Sun drives the Earth's climate, the exact physical mechanisms by which solar variations manifest themselves in the climate system are yet poorly understood. One reason is that it's hard to separate each solar driver, e.g. UV, TSI and cosmic rays, and pinpoint it to a single climate response. Table 18.1 gives a summary of the most often cited mechanisms.

Table 18.1 The pathways through which the climate of the lower atmosphere is influenced by solar variability

Forcing	Mechanism
Solar UV irradiance	Heating the middle and upper atmosphere, dynamical coupling down to troposphere
	Lower and middle atmosphere composition and chemistry; impacts radiative forcing and temperature structure
Total solar irradiance (variations due to variable solar emission or orbital variations)	Radiative forcing of climate
	Direct impact on hydrological cycle and sea surface temperatures
Galactic cosmic rays	Impact on condensation nuclei, impact on electric field and ionisation of lower atmosphere
Solar energetic particles	Impact on temperatures and composition
	Ionosphere–magnetosphere–thermosphere coupling
	Ionisation of middle and upper atmosphere

18.2   Other Influences, e.g. Galactic Cosmic Rays

Apart from direct solar drivers, galactic cosmic rays (GCR) have also been suggested to influence the climate of the Earth. During high solar activity, the amount of GCRs coming from outer space is reduced. Thus, fewer GCRs during the active solar phase are observed as shown in Fig. 18.2. Galactic cosmic rays have impact on ionisation of the lower atmosphere and the global electric field. It subsequently modulates cloud amount triggering cloud condensation nuclei (Tinsley 1996).

18.3   Sunspot vs. Galactic Cosmic Ray (GCR)

Figure 18.3 shows sunspot number, and GCR is strongly anti-correlated. It is because, during active solar years when there are more sunspot numbers, the GCR have difficulties in penetrating the stronger shield of high-energy particles. As SSN and GCR are strongly anti-correlated, SSN can also be used as a proxy for GCR for its climate influence.

To summarise, part III discusses on other major influences on climate. It described possible limitations in attributing and identifying the role of the sun. Later, it covered a discussion on Arctic and Antarctic Sea ice, which was followed by exploring CMIP5 project and some results. In the subsequent chapter, it focused on greenhouse effect. Absorption of IR radiation in various wavelengths by water vapour and

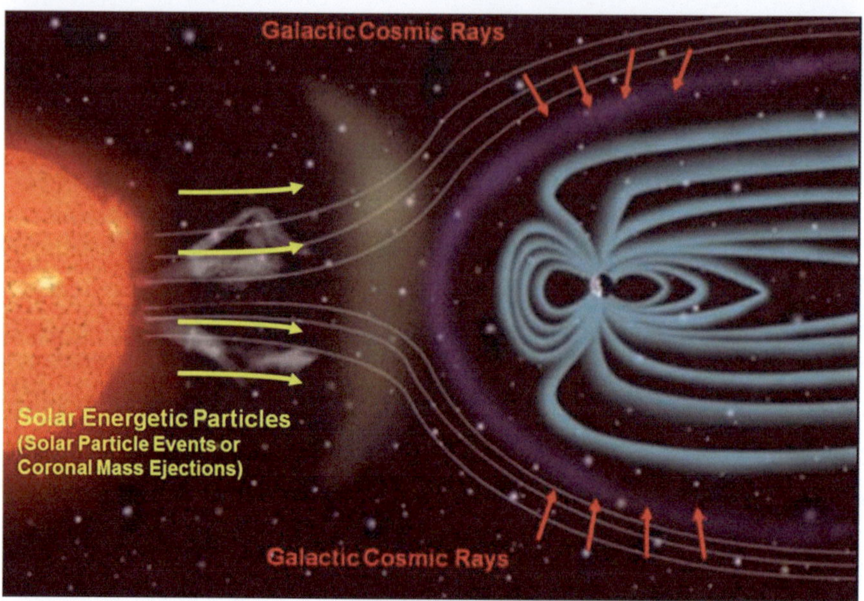

Fig. 18.2 Galactic cosmic rays shown by red are reduced during active solar years when more solar energetic particles hit the Earth's atmosphere (shown by yellow colour). (Source: https://commons.wikimedia.org/wiki/File%3APIA16938-RadiationSources-InterplanetarySpace.jpg, uploaded on 30/12/17, picture by NASA/JPL-Caltech/SwRI (http://photojournal.jpl.nasa.gov/jpeg/PIA16938.jpg) [Public domain], via Wikimedia Commons)

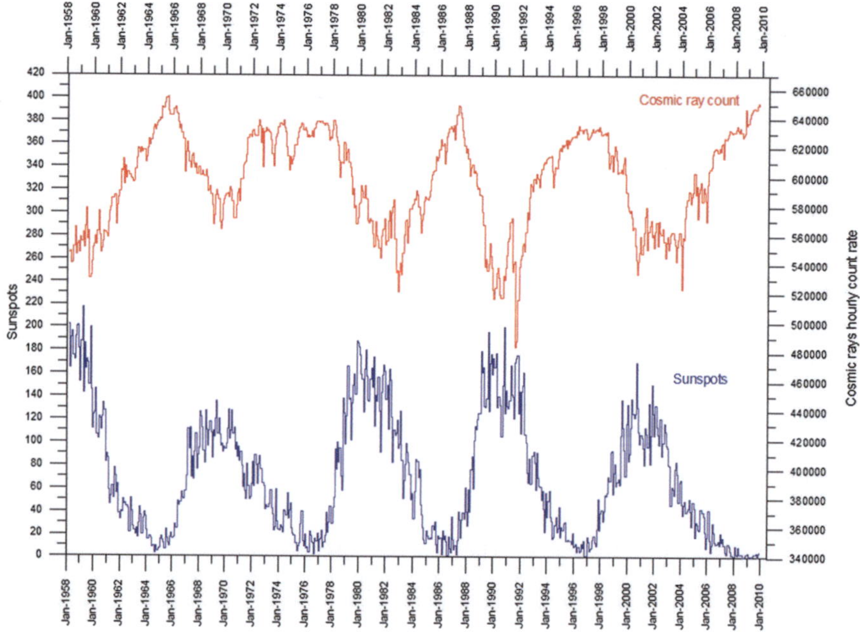

Fig. 18.3 Time series of sunspot number (blue) and GCR (red). (Source: http://www.climate4you.com/Sun.htm# Cosmic ray intensity and sunspot activity, uploaded on 30/12/2017, credit Germany Cosmic Ray Monitor in Kiel (GCRM) and NOAA's National Geophysical Data Center (NGDC))

CO2 is shown. Radiative forcing as presented in AR4 (IPCC 2007) and AR5 (IPCC 2013) are reviewed. It discussed years of major volcanos and various effects on climate. Antarctic ozone hole and banning of CFCs through Montreal Protocol are also addressed. It was shown that ozone in the stratosphere and greenhouse gas in the troposphere both have a comparable effect on surface climate, though ozone is more important in higher latitudes and altitudes. Apart from solar 11-year cyclic variability, other solar outputs are also important in modulating the climate of the earth and discussed briefly.

References

Baumgaertner AJG, Seppälä A, Jöckel P, Clilverd MA (2011) Geomagnetic activity related NOx enhancements and polar sur-face air temperature variability in a chemistry climate model: modulation of the NAM index. Atmos Chem Phys 11:4521–4531. https://doi.org/10.5194/acp-11-4521-2011

Bochnıcek J, Hejda P (2005) The winter NAO pattern changes in association with solar and geomagnetic activity. J Atmos Sol Terr Phys 67:17–32. https://doi.org/10.1016/j.jastp.2004.07.014

Bochnıcek J, Davıdkovová H, Hejda P, Huth R (2012) Circulation changes in the winter lower atmosphere and long-lasting solar/geomagnetic activity. Ann Geophys 30:1719–1726. https://doi.org/10.5194/angeo-30-1719-2012

Funke B, López-Puertas M, Gil-López S, von Clarmann T, Stiller GP, Fischer H, Kellmann S (2005) Downward transport of upper atmospheric NOx into the polar stratosphere and lower mesosphere during the Antarctic 2003 and Arctic 2002/2003 winters. J Geophys Res 110:D24308. https://doi.org/10.1029/2005JD006463

Gray LJ et al (2010) Solar influences on climate. Rev Geophys 48:RG4001. https://doi.org/10.1029/2009RG000282

IPCC (2007) Fourth assessment report of the intergovernmental panel on climate change. Cambridge University Press, Cambridge/New York

IPCC, Climate Change (2013) The physical science basis. Contribution of working group I to the fifth assessment report of the intergovernmental panel on climate change. Cambridge University Press, Cambridge/New York. http://www.climatechange2013.org/images/report/WG1AR5_ALL_FINAL.pdf

Maliniemi V, Asikainen T, Mursula K, Seppälä A (2013) QBO-dependent relation between electron precipitation and wintertime surface temperature. J Geophys Res Atmos 118. https://doi.org/10.1002/jgrd.50518

Maliniemi V, Asikainen T, Mursula K (2014) Spatial distribution of Northern Hemisphere winter temperatures during different phases of the solar cycle. J Geophys Res Atmos 119. https://doi.org/10.1002/2013JD021343

Roy I (2014) The role of the Sun in atmosphere-ocean coupling. Int J Climatol 34(3):655–677

Roy I (2018) Solar cyclic variability can modulate winter Arctic climate. Sci Rep 8(1):4864. https://doi.org/10.1038/s41598-018-22854-0

Roy I, Collins M (2015) On identifying the role of Sun and the El Niño Southern Oscillation on Indian Summer Monsoon Rainfall. Atmos Sci Lett 16 (2):162–169

Roy I, Asikainen T, Maliniemi V, Mursula K (2016) Comparing the Influence sunspot activity and geomagnetic activity on surface climate. J Atmos Sol Terr Phys. https://doi.org/10.1016/j.jastp.2016.04.009

Rozanov E, Callis L, Schlesinger M, Yang F, Andronova N, Zubov V (2005) Atmospheric response to NOy source due to energetic electron precipitation. Geophys Res Lett 32:L14811. https://doi.org/10.1029/2005GL023041

Seppälä A, Verronen PT, Clilverd MA, Randall CE, Tamminen J, Sofieva V, Backman L, Kyrölä
    E (2007) Arctic and Antarctic polar winter NOx and energetic particle precipitation in 2002-
    2006. Geophys Res Lett 34:L12810. https://doi.org/10.1029/2007GL029733
Sinnhuber M, Kazeminejad S, Wissing JM (2011) Interannual variation of NOx from the lower
    thermosphere to the upper stratosphere in the years 1991-2005. J Geophys Res 116:A02312.
    https://doi.org/10.1029/2010JA015825
Tinsley BA (1996) Correlations of atmospheric dynamics with solar wind-induced changes of air-
    earth current density into cloud tops. J Geophys Res 101:29,701–29,714

Few Questions and Exercises for Students

1. Define Hadley circulation, Walker circulation, Ferrel cell and Polar cell.
2. Define ENSO, PDO, AMO, NAO, AO, AAO, QBO, SSW and IOD.
3. Give a brief description about stratosphere-troposphere coupling.
4. What is the difference between PDO and ENSO?
5. What is the difference between NAO and AMO?
6. Discuss the mechanism of QBO formation.
7. For SSW, what is the difference between major warming, minor warming and final warming?
8. Give brief descriptions about solar 'top-down mechanism' and 'bottom-up mechanism'.
9. What is Indian summer monsoon?
10. Define greenhouse effect.
11. What is the role of volcanos on climate?
12. Discuss the stratospheric polar temperature in various combinations of sun-QBO phases.
13. What do you mean by solar peak years compositing?
14. What do you mean by solar minimum year compositing?
15. Discuss the mechanism how solar 11-year cyclic variability influences polar vortex and tropical lower stratosphere.
16. How polar vortex is influenced by ENSO or QBO?
17. Discuss ENSO teleconnections in various seasons.
18. Discuss different Nino index and how those are constructed?
19. What do you mean by Canonical and Modoki ENSO? Why do we need to study those types of ENSO separately?
20. What is TSI? Can you name few TSI reconstructions and tell briefly how those are constructed?
21. What are ocean conveyor belt and thermocline?
22. Discuss the mechanism of ENSO interannual variability involving oceanic Rossby and Kelvin waves.

I. Roy, *Climate Variability and Sunspot Activity*, Springer Atmospheric Sciences, https://doi.org/10.1007/978-3-319-77107-6

23. Discuss studies that show the connection between the sun and ENSO in the later part of twentieth century.
24. What was shown by research about the sun-ENSO connection in the earlier period?
25. Can you show how two similar studies can show a contradiction in terms of sun-QBO connections?
26. What is the difference between multiple linear regression technique and compositing technique to detect a solar signal?
27. Where is the location of Aleutian Low (AL)? Can you tell few studies that detected solar signal around AL and what is the nature of those signals?
28. How solar influence around pole changes when QBO is also taken into consideration?
29. What is the role of the thermocline in changing phases of ENSO?
30. What are CMIP5 models?
31. Tell few areas where CMIP5 do not agree with observations.
32. Tell few areas where there are agreements.
33. What are historical and RCP scenarios?
34. Define radiative forcing.
35. Discuss ozone depletion in the stratosphere.
36. Some data sources are mentioned at the end of section 11 (before summary). Can you tell what the correlation between SSN and ENSO is? Use different 50 years period and show if there are any changes.
37. Find the correlation coefficient between QBO and ENSO, and QBO and SSN. Use different time periods.
38. Use KNMI Climate Explorer site (https://climexp.knmi.nl/selectindex.cgi?id=someone@somewhere). Take various monthly climate index, e.g., NAO, AO, IOD, AAO. Do similar analyses as mentioned in 36 and 37.
39. Do similar studies using various regression techniques.
40. Repeat those using compositing techniques.
41. Are the results different if you take different seasons for, e.g., Dec-Jan-Feb (DJF), Mar-April-May (MAM), June-July-Aug (JJA) and Sept-Oct-Nov (SON)?

Further Reading

Part I

Brönnimann S, Ewen T, Griesser T, Jenne R (2006) Multidecadal signal of solar variability in the upper troposphere during the 20th century. Space Sci Rev 125(1–4):305–317. https://doi.org/10.1007/s11214-006-9065-2

Fleming EL, Chandra S, Barnett JJ, Corney M (1990) Zonal mean temperature, pressure, zonal wind, and geopotential height as functions of latitude. Adv Space Res 10:11–59

Plumb RA (1984) The quasi-biennial oscillation. In: Holton JR, Matsuno T (eds) Dynamics of the middle atmosphere. Terra Science, Tokyo, pp 217–251

Randel WJ et al (2004) Changes in column ozone correlated with the stratospheric EP flux. J Meteorol Soc Jpn 80(4B):849–862. https://doi.org/10.2151/jmsj.80.849

Roy I, Haigh JD (2010) Solar cycle signals in sea level pressure and sea surface temperature. Atmos Chem Phys 10(6):3147–3153

Simpson IR, Blackburn M, Haigh JD (2009) The role of Eddies in driving the tropospheric response to stratospheric heating perturbations. J Atm Sci 66(5):1347–1365

Weng H (2005) The influence of the 11-yr solar cycle on the interannual-centennial climate variability. J Atmos Sol-Terr Phys 67:793–805

Part II

Dima M, Lohmann G, Dima I (2005) Solar-induced and internal climate variability at decadal time scales. Int J Climatol 25:713–733. https://doi.org/10.1002/joc.1156

Roy I, Haigh JD (2011) The influence of solar variability and the quasi-biennial oscillation on lower atmospheric temperatures and sea level pressure. Atmos Chem Phys 11:11679–11687. https://doi.org/10.5194/acp-11-11679-2011

Solanki SK, Fligge M (1998) Solar irradiance since 1874 revisited. Geophys Res Lett 25(3):341–344

Solanki SK, Krivova NA (2003) Can solar variability explain global warming since 1970? J Geophys Res 108(A5):1200. https://doi.org/10.1029/2002JA009753

Part III

Baumgaertner AJG, Seppälä A, Jöckel P, Clilverd MA (2011) Geomagnetic activity related NOx enhancements and polar sur-face air temperature variability in a chemistry climate model: Modulation of the NAM index. Atmos Chem Phys 11:4521–4531. https://doi.org/10.5194/acp-11-4521-2011

Haigh (2007) The Sun and the Earth's climate, living. Rev Solar Phys 4:2

Krishnamurthy V, Kinter JL (2002) The Indian monsoon and its relation to global climate variability. In: Rodo X, Comin FA (eds) Global climate – current research and uncertainties in the climate system. Springer, Berlin, pp 186–236

Lüning S, Vahrenholt F (2017) Front Earth Sci 12 December 2017 | https://doi.org/10.3389/feart.2017.00104

Roy I (2018) Solar cyclic variability can modulate winter Arctic climate. Sci Rep 8(1):4864. https://doi.org/10.1038/s41598-018-22854-0

Sinnhuber M, Kazeminejad S, Wissing JM (2011) Interannual variation of NOx from the lower thermosphere to the upper stratosphere in the years 1991–2005. J Geophys Res 116:A02312. https://doi.org/10.1029/2010JA015825

Index

FSC
www.fsc.org
MIX
Papier aus verantwortungsvollen Quellen
Paper from responsible sources
FSC® C105338